Optical Fiber Systems: Technology, Design, and Applications

Optical Fiber Systems: Technology, Design, and Applications

CHARLES K. KAO

Vice President and
Director of Engineering
Electro-Optical Products Division
ITT

McGraw-Hill Book Company

New York · St. Louis · San Francisco · Auckland
Bogotá · Hamburg · Johannesburg · London · Madrid
Mexico · Montreal · New Delhi · Panama · Paris
São Paulo · Singapore · Sydney · Tokyo · Toronto

Library of Congress Cataloging in Publication Data

Kao, Charles K., date.
  Optical fiber systems.

  Includes index.
  1.Fiber optics.   I.Title.
TA1800.K36   621.36′92   81-14300
                                     AACR2

Copyright © 1982 by McGraw-Hill, Inc. All rights reserved. Printed in the United States of America. Except as permitted under the United States Copyright Act of 1976, no part of this publication may be reproduced or distributed in any form or by any means, or stored in a data base or retrieval system, without the prior written permission of the publisher.

1234567890   KPKP   8987654321

ISBN 0-07-033277-0

The editors for this book were Barry Richman and James T. Halston, the designer was Jules Perlmutter, and the production supervisor was Thomas G. Kowalczyk. It was set in Vermilion by The Kingsport Press.

Printed and bound by The Kingsport Press

Contents

	Preface	ix
	Abbreviations	xi
1	**The Role of Optical Fiber Systems**	1
	Optical Fiber Systems in Sociological Evolution	2
	Characteristics of Optical Fiber Systems	9
2	**Algorithm for Optical Fiber Link Design**	15
	Introduction	15
	Design Algorithm	18
3	**Fiber**	21
	Introduction	21
	Physical Structure	22
	Basic Theory	22
	Comparison of Waveguide Types	32
	Fiber Imperfections	37
	Fiber Packaging	41
	Physical Properties of Optical Fiber Waveguides	44
	Fiber Evaluation Methods	55
	Measurements of Optical Characteristics of Fiber Waveguides	56
	Measurement of Mechanical Characteristics of Fiber Waveguides	59
	Fiber-Fabrication Processes	62
	Fiber-Drawing Processes	69

4 Fiber Cables — 75

Strength Members — 76
Cable Structures — 77
Cable Testing — 80

5 Light Sources — 83

Types of Light Sources — 84
Basic Characteristics of Light Sources for Communications — 86
Selection of a Light Source — 90

6 Modulation and Detection — 103

Introduction — 103
Photodetectors — 107
Receiver Design Considerations — 112

7 Fiber Connectors, Splices, and Couplers — 117

Introduction — 117
End Preparation — 120
Optical Fiber Splicing — 123
Optical Fiber Connectors — 125
Couplers — 129

8 Systems — 137

Interoffice Link — 138
Entrance Links — 139
24 or 30 Channel PCM Link for a High-Electromagnetic-Radiation Field Environment — 140
Undersea Long-Haul Systems — 140
Cable TV Trunking — 143
Tamperproof Link — 144
Military Applications — 145
Mobile Link — 145
Radar Remote — 147
Weapon Guidance — 147
Torpedo Guidance — 149
Tethered Vehicles — 150
Distribution Systems — 150
Wired Office — 151
Wired City — 151

	Cable TV Distribution	156
	Sensor Systems	157
9	**Anatomy of a Design**	**159**
	Introduction	159
	Section 1 Link Design	163
	Section 2 Link Design	175
10	**System Economics**	**183**
	First-Stage Considerations	184
	Second-Stage Considerations	185
	Third-Stage Considerations	187
	Fourth-Stage Considerations	192
	Fifth-Stage Considerations	194
	Selected Bibliography	**197**
	Index	**199**

Preface

Technology is always socially relevant. Too often we allow ourselves to be mesmerized by the wonders of technology and devote our energy to its advancement without the necessary reflection on its social relevance. This practice could lead to incomplete realization or inappropriate use of the full potential of a technology. In optical fiber system technology we are dealing with one of the cornerstones of our information age in which the efficient utilization of information governs our well-being. The understanding of its sociological implications is extremely important.

With this in mind, this book is written to give the readers a glimpse of the range of application and the impact of this technology, as well as to introduce the associated technology of optical fiber systems in a concise and, hopefully, clear manner. The reader will be able to see the interrelationship of the technology and its implications and will be able to construct the basis of a system design and perceive the relative advantages and disadvantages.

Chapter 1 delineates the roles of optical fiber systems from a sociological evolutionary point of view. It introduces the major characteristics of such systems. Chapter 2 gives the design rules but avoids explicit discussion of the design approaches. Chapters 3–7 introduce the key components and their associated background technology. The discussions aim at providing the readers with an understanding of the interrelation of the salient features from a system point of view. References are given for readers to access complete theoretical derivations. These chapters cover fiber, cable, sources, detectors, and couplers. Chapter 8 provides a view of a selected number of systems for a variety of applications. These examples are chosen to highlight where optical fiber systems offer advantages. Chapter 9 illustrates the design process via a specific example. It shows how the knowledge of different components is needed to achieve a viable design. It brings the algorithm outlined in Chapter 2 into focus. Chapter 10 looks at the system economics to bring the book to a close.

This is a textbook but not a source book. It complements other books written more explicitly as textbooks as well as special treatises, so that the reader can get broader viewpoints and finer insights.

Over the period this book was in preparation, I was helped by my wife, son, and daughter, who gave me the understanding and support needed. I am indebted to the various secretaries who typed the manuscript at different stages, particularly Mrs. Havens, who typed and compiled the final manuscript. I would like to thank my colleagues at ITT and other colleagues in the optical fiber field who allowed me to refer to their work and gave me much valuable assistance. Last, but not least, I would like to express my gratitude to a great number of scientists, engineers, managers, and people from many walks of life who have and who are contributing to this technology, for without them I would not be able to write this book.

Charles K. Kao

Abbreviations

AGC	automatic gain control	**MTBF**	mean time before failure
APD	avalanche photodiode	**MTF**	modulation transfer function
BER	bit-error rate	**NA**	numerical aperture
CCIR	International Radio Consultative Committee	**NRZ**	nonreturn-to-zero
		NTSC	National Television Standards Committee
CCITT	International Telegraph and Telephone Consultative Committee	**OD**	outer diameter
		PCM	pulse-code modulation
CCTV	closed-circuit TV	**PIN**	positive-intrinsic-negative
CVD	chemical vapor disposition	**PSK**	phase-shift keying
CW	continuous-wave	**PTT**	post, telephone, and telegraph
ECL	emitter-coupled logic		
EMI	extromagnetic interference	**PVAC**	present value annual cost
FCC	Federal Communications Commission	**RF**	radio frequency
		RMS	root-mean-square
FDM	frequency-division multiplex	**RTV**	room temperature vulcanization
FET	field-effect transistor		
FSK	frequency-shift keying	**RZ**	return-to-zero
IF	intermediate frequency	**SNR**	signal-to-noise ratio
IR	infrared	**TDM**	time-division multiplex
LED	light-emitting diode	**TTL**	transistor-transistor logic
LSI	large-scale integration	**VLSI**	very large scale integration

Chapter

The Role of Optical Fiber Systems

Optical fiber waveguides are vital in an "information society," where the primary preoccupation in life is generation, dissemination, and management of information. In such a society we can readily appreciate that job efficiency, productivity, and meaningfulness of our endeavors are improved by a better organization and control of our information resources. Our ability to transport information efficiently is highly relevant to our well-being.

An optical waveguide is a threadlike structure capable of handling the transportation of a large volume of information traffic. We need it as the building block of our information highway system to help us in managing our energy resources, transportation, and communications; delivering health care and community services; strengthening our military defense; developing business; and providing materials for our entertainment and education. Optical waveguides are destined to find a myriad of applications in a ubiquitous role.

The move from our material-based society into an information-based society is prompted by economic and technological developments. The growth of population has reached a point where our survival is based increasingly on a complex set of criteria. We need to extend further our physical and mental capabilities in order to cope with increasing production needs and decreasing energy availability. In recent years our technological advances have provided us with three important innovations which enable us to develop tools to tackle the tasks at hand. The invention of the transistor and its evolution into the integrated circuits allow densely packaged, complex, and powerful electronic circuits to be designed. These are required to extend our capability to implement complex control functions, provide large information storage, and exercise fast information processing. In particular, a full range of computers is now available to extend our mental capability. The invention of lasers enables us to utilize the optical wavelength range of the electromagnetic energy spectrum. The optical lasers are equivalent radio-frequency (RF) oscillators and can serve as information carriers. Because of their high frequency ($\sim 3 \times 10^{14}$ Hz), they can theoretically accommodate many orders of magnitude more bandwidth of information than can an RF information carrier. The invention of the optical fiber waveguide provides us with the transmission media par excellence. The transportation and distribution of information at a high rate becomes feasible.

Optical Fiber Systems in Sociological Evolution

The physical and mental needs of the people in an information society are both materialistic and service-oriented. The materialistic aspects refer to the basic production of food and the conversion of raw materials into general merchandise. The service aspects require a new industry to improve the efficiency of the distribution of material goods, balance the resource allocation, and enrich life with new pursuits. At the basis of this service industry are optical fiber systems forming the communications links.

ENERGY RESOURCES

Until the control of thermal nuclear fusion can be perfected in the twenty-first century, available basic energy resources are limited. Many of the readily available resources such as oil and gas are rapidly being depleted. As a result, energy cost is increasing rapidly. This calls for a careful examination of the mode of energy usage, with a view to fulfill the increasing demand more efficiently and safely. It is a process in which the consumption of oil, natural gas, coal, solar energy, and other resources for domestic and industrial activities is balanced for cost and availability. It is time to trade heavy energy-expending activities, such as the making of articles for rapid consumption, for the manufacture of more durable goods. It is time to explore alternative primary energy resources while taking precaution against undesirable environmental impacts.

Here the fiber system plays a minor but important role. The characteristic of fiber—of immunity to electromagnetic interference—allows fiber systems to be used with advantages in electrical generation and transformer stations to improve control and communications functions.

TRANSPORTATION

The physical movement of people and goods is an essential part of life, and different transportation requirements are met by different means. For mass transit and bulk goods movement, there are buses, trains, ships, and large trucks. For individual journeys, there are automobiles and bicycles. For rapid transit, there are airplanes. Together, these methods of transportation keep the society functioning.

The volume of traffic has been expanding rapidly as population and trade increase. The total expenditure of energy for transportation has reached a level which is sufficiently large, compared with the level of energy reserve, to warrant attention. More efficient use of the transportation system is important.

The role of fiber systems in transportation is indirect. In mass transit systems the operation can be made much more efficient with the provision of reliable means of control. This usually involves locating the position and speed of the vehicles on the railway. Along the high-speed electrified railway line, such a need is most obvious. Fiber control systems can be installed with advantage along the electrified track where copper wire systems would encounter serious electromagnetic interference problems.

COMMUNICATIONS

The introduction of optical fiber systems will revolutionize the communications network. The low-transmission loss and the large-bandwidth capability of the fiber systems allow signals to be transmitted for establishing communications contacts over large distances with few or no provisions of intermediate amplification. Also, more information can be transmitted over a shorter time than with the use of alternative systems. This means that the optical fiber communications network can provide more services at a lower cost.

The telephone network, based on copper wires as the transmission lines, has provided us with a very valuable communications network. The instant audio contact through a telephone has increased business efficiency, reduced the need to travel, and provided many services which otherwise would not be available.

The communications network, based on optical fiber as the transmission lines, has several orders of magnitude more information-carrying capacity per unit time than does the telephone network and can provide a host of services, including video services, which require a bandwidth much greater than that used in the telephone service. Such a communications network allows video, audio, and data transmission, and in association with computers, it literally allows our visual senses and certain of our brain functions to be extended. The power of such a network would create a completely new sociological environment—one which befits the information age.

The evolution toward such a communications network is likely to be relatively slow and deliberate. The existing network represents a huge monetary investment and cannot be written off overnight. Neither can this investment of an even larger sum of money to create the new network be done instantaneously. There will be a gradual evolution.

The advantages of a fiber system can be exploited most readily in the existing network in a number of important areas. In metropolitan areas copper cables are installed mainly in underground conduits, providing connection between busy exchange offices. Fiber cables with equal traffic-carrying capacity are much smaller in size and can provide interexchange office connection without intermediate repeaters in most cases. The lifetime system cost—which includes installation, maintenance, and hardware costs—favors fiber systems, especially where traffic density and growth are high. Intercity routes with high traffic densities can also be served better and more economically by fiber systems. These are the areas where fiber systems are being introduced into the existing network. The network between a satellite ground receiving station and a distribution hub provides another opportunity for early fiber system entry. Trunking of television signals over distances of several kilometers and upward can be most attractive on fiber.

The increased usage of fiber systems will result in reduction of their cost, when the economy of scale takes effect. Fiber systems will then be preferred on the basis of cost alone, even in areas where their advantages of large bandwidth and low loss are not needed. Early introduction of fiber cable into the subscriber distribution network will then be possible, thus laying the foundation of the wide-bandwidth information network of the future.

Information services for the home and business premises already being used include document transmission (facsimile); telex; 1200- to 9600-baud data for computerized banking; airline, hotel, and other reservations; dial-a-message service on telephones for weather, local events, and other information; and stored messages from special customer services. Some prospective services being tested include package switching networks for improved data services and computer access, video text for stored information such as restaurant guides, train timetables, programmed teaching, or, for more specific customers, stock market information, inventory control, and so on, and for improved security and energy management, remote alarm and meter monitors. Services envisaged for the future usually involve the use of increased bandwidth to provide faster information transfer and the use of live video information in an interactive manner. While most of the services can be provided without the need of a broadband transmission medium, the increased use of information would result in broadband transmission in an increasingly larger area of the communications network. Eventually with the widespread use of broadband services, the capability of broadband distribution to individual subscribers would be required.

Fiber cost is expected to decrease substantially to a level where its use in the distribution network becomes non-cost-prohibitive. Its use is expected to be very widespread, and it will lay the foundation for the present network to evolve into the future broadband network.

HEALTH SERVICES

Health care delivery requires individual consultation by physicians and massive specialist hospitals with sophisticated equipment. Communications both within large hospitals and to and from these establishments is vital for operational efficiency and effectiveness. If a physician can call for the relevant records of a patient, vital statistics, rapid analyses, and a glossary of information aids at the hospital or at the home of a patient, the health care delivery would be vastly more effective.

Fiber systems can provide the communications links capable of handling visual instruments and fast computer-controlled data equipment without suffering from electromagnetic interferences which are often associated with high-powered hospital equipment, such as x-ray machines. The visual observation aids are particularly important. Remote monitoring of patients and recalling of specialized symptomatic information and surgical procedures in visual form can bring the massive resources of a specialist hospital to the assistance of a physician in a remote village.

In a role rather different from that of an information transmission line, an optical fiber as a conductor of light and visual images is exploited in medical instruments for illumination and observation of inaccessible areas. As transmission along a fiber is better controlled, fibers capable of transmitting images in real-time holographic mode can be envisaged. Single-fiber systems can be designed to deliver a precise amount of optical power for pathological analysis and as a surgical tool or as new means of curing certain diseases such as the control and destruction of tumors.

COMMUNITY SERVICES

The wired-city concept has been proposed in which a community is intimately wired together in such a manner that a host of community-related facilities such as local news, shopping guide, neighborhood festivities, community library, rescue squad, and a computer bank are available to the community as a whole.

Experimental system has shown that the value of a wired city is associated with the perception of the people. It is a sociological experiment with far-reaching implications, but the results so far are too influenced by the main social trends to be really meaningful. Even in this context, wired-city development can be seen to promote a new community spirit and a different way of utilizing services.

Fiber systems can and will be the leading contender as the principal transmission media for wired cities. The provision of integrated services over a wide-bandwidth, low-loss transmission system is a desirable solution.

MILITARY DEFENSE

The special features of an optical fiber are small size, light weight, strength, flexibility, wide temperature range, and interference freedom, in addition to wide bandwidth and low loss. These features are key to improving the strategic and tactical capabilities of the military forces.

More powerful communications networks can be created and installed under various conditions. In the strategic base communications application, the more compact cable enables the cables to be easily transported and permits a variety of permanent and mobile configurations to be implemented easily. The remote connection to radar sites from the signal processing station can be rapidly deployed, and the wide bandwidth and the low loss allow longer spacing between the radar site and the station, thus allowing a greater safety for the operating personnel. The electromagnetic interference freedom characteristics can be used with great advantage on ships, airplanes, and armored vehicles, where much data are being processed under electrically noisy environment. The interference freedom also allows the system to preserve a high degree of privacy. This is utilized in systems where sensitive data are to be transmitted. The high strength and the flexibility allow fiber systems to be envisaged for tactical applications. Wire-guided weapons can be made to cover a longer range and with more precise guidance or even with visual target search capabilities. Fibers can also improve the capabilities of towed surveillance vehicles by improving the information and mechanical performance of the towing cable. The propagation characteristics of the fiber can be utilized to indicate strain and temperature change. This can be considered for sensor applications with improved sensitivity. Acoustic and magnetic field sensors and a fiber gyroscope have been identified as possibilities.

BUSINESS DEVELOPMENT

Akin to the wired-city concept is the "office-of-the-future" concept, where business functions of information retrieval, distribution, dissemination, and analysis are performed by equipment available within the office of the future. A general manager can call for financial and operational data with the touch of a button or a simple voice command. A document can be typed into a word-processing machine to allow for easy editing. Subsequent production of personalized versions can be automatically routed internally, thus enabling staff in other parts of the office to access the document on their terminal equipment in a display or hard-copy form and externally, for remote distribution. Inventory control, numerical control of machines and payroll preparation are just a few of the many automated services which an office of the future may have. Audiovisual conferencing and interactive graphics are other possible features.

The technical realization of such a system is not too different from the wired city, but since the system may be confined to a single building, and since the community involved has a more homogeneous and identifiable requirement, the system requirements can be more readily met.

ENTERTAINMENT

The pursuit of happiness often starts with a satisfying job, followed by spending the affordable surplus earnings on entertainment. We are lured to a world of service industry aimed at entertainment. The spectator sports, the stage, the movies, the radio, and television (TV) programs, as well as toys and books, are just a handful of more pertinent examples.

Perhaps TV has emerged as the most influential of all entertainment media. Backed by revenue from advertising, television broadcasters have captured the attention of millions of people each day and entertain them with a great variety of visual programs. It has dramatically altered the proportion of active participants in entertainment to passive spectators.

Recently, cable television [also referred to as *community antenna television* (CATV)] through the use of geosynchronous communication satellites, is changing the TV industry. Cable TV provides many channels in each house, giving viewers a greater degree of freedom of choice of programs. The commercialization of video tape recorders and video disks is another entry into the TV industry that is destined to reshape the mode of operation further.

The entertainment market accepts many service vendors who compete for the surplus earnings of the consumers by persuading them to become their customers, promising them entertainment which is worth much more than the payment. For example, cable TV charges a flat monthly fee for the right to have more programs to choose from, while pay TV charges an extra fee for the right to see mainly movies with no commercials. The charge is low in comparison to the cost of going to a movie at a cinema. The video disk merchants are appealing to the consumer's pride of ownership and individual choice, while video recorder merchants are appealing to the consumer's desire to view scheduled programs at the most convenient time. In the meantime, the cinema, the stage, and books continue to flourish but cater to a more select group.

The possibility of providing broadband information delivery to each home with an optical fiber network and the further possibility of delivering video services via that network on a switched basis are important in shaping the future of the entertainment world. The switched TV network can provide programs of wide audience appeal to cater to the majority while serving programs of limited appeal to more selected audiences by suitably arranging a revenue base compatible with a narrow business basis. Such a scheme resembles a video library and provides the ultimate flexibility in individual choice.

EDUCATION

With the technology available, the education process could be made more effective. For instance, the availability of pocket tape recorders at a certain university prompted some students to skip classes and ask friends to place their recorders at the class to record the lecture. This continued for some time, and the popularity of such a practice grew until one day a few students walked into the classroom and found it full of recorders but with the professor absent. His recorder was there with the "play" button down. This story has more than one moral. One of the morals prompts the question, "Is a prerecorded lecture better or worse?"

In fact, with computers and video tape, a prepared teaching tape instead of a live lecture together with the use of computer-aided instruction can be most effective. The role of the professor then becomes that of the mentor who leads live sessions of debates and criticisms and opens the minds of the young by pitting them against the mind of the experienced.

Already educational TV and program learning on computers are being used in schools and universities. The successful Open University in the United Kingdom (at Walton, Buckinghamshire, England) demonstrated convincingly the power of TV as an educational tool. Universities with large student bodies in several campuses found closed-circuit television (CCTV) to be an effective means of making courses available. In fact, prepared lecture series are appearing as educational commodities. In due course authoritative texts should be used to strengthen poor teaching, and much more information should be made accessible as references at information banks.

The provision of broadband communications systems in educational institutions is a definite necessity. The role of fiber optics for this application is very clear.

Characteristics of Optical Fiber Systems

BASIC LINK

The basic optical fiber system is illustrated in Fig. 1-1. It consists of a transmitter which transforms an electrical signal to be transmitted into an optical signal, a receiver which converts the optical signal back to the original electrical form, and a fiber transmission line which conducts the optical signal from the transmitter to the receiver.

Three new components are involved: the light source, the photodetec-

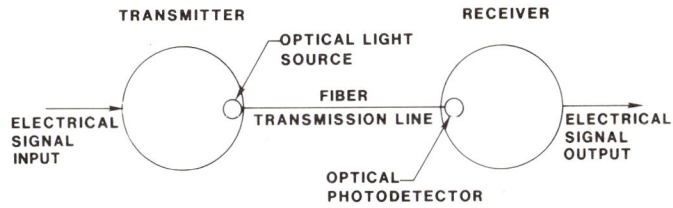

Fig. 1-1 A basic optical fiber system.

tor, and the optical fiber transmission line. The optical light source generates the optical energy which serves as the information carrier, similar to a radio-wave source supplying electromagnetic energy at radio-wave wavelengths as the information carrier. The optical photodetector detects the optical energy and converts it into an electrical form. The optical fiber transmission line is the equivalent of a pair of copper wires and functions as the conductor of optical energy.

The fiber transmission line must be suitably packaged and incorporated into the cable structure to withstand the forces involved during and after installation and to survive over extended periods in different environments. The fibers and cables must also be joined by splicing forming permanent joints or via connectors which allow repeated connections. For monitoring and distribution purposes, different couplers are also required.

The major characteristics of optical fiber waveguides that distinguish optical fiber from other systems can be separated into three categories: physical, optical, and special characteristics. These characteristics are briefly introduced prior to more detailed discussion later.

PHYSICAL CHARACTERISTICS OF OPTICAL FIBER WAVEGUIDES

Optical fiber waveguides are threadlike structures made from dielectric materials in a glassy form. A typical fiber has a diameter of 125 μm. This is about the thickness of human hair. It can be readily made into long lengths with uniform dimensions. Unit lengths of several kilometers are common, and much longer lengths are possible.

Optical fiber waveguides for operation in the 0.5- to 1.6-μm wavelength region are generally made with inorganic oxide glasses with a high silica content. Such waveguides have a specific gravity of about 2.3. Since the volume of a 125-μm-diameter fiber 1 km in length is about 12 cm^3 and weighs about 28 g, it is very small in size and light in weight. Since the volume of material is small, the intrinsic cost is low, especially because the materials involved are not rare elements. Fiber cost is made up of the basic material portion and the labor portion incurred in the fabrication process. Fabrication methods for low-loss fiber involves the use of high-purity materials, and in one process the fiber material is made at the rate of around 1 to 2 g/min. As volume of production increases, the fiber cost will decrease until it reaches an asymptotic value for a particular fabrication process. The eventual cost is expected to be comparable even to a pair of 22-gauge copper wires.

Since glass is intrinsically a very strong and durable material, properly made and protected fiber is strong and durable. It is also highly flexible and can negotiate bends as small as a few millimeters. For short durations, fiber can be bent to a radius of 1 to 2 mm without breaking and will return to its original state on account of its staying perfectly elastic up to more than 10 percent elongation.

OPTICAL CHARACTERISTICS OF OPTICAL FIBER WAVEGUIDES

The optical transmission characteristics of the optical fiber waveguides are expressed in terms of attenuation or loss and bandwidth or pulse dispersion. The attenuation causes optical energy to be dissipated along the waveguide during transmission and reduces the available energy at the destination after transmitting along the length of the fiber. Since the transmitter power output and the detector sensitivity are fixed for a given signal and operating condition, the fiber attenuation governs the maximum path length that the signal can span without amplification. The bandwidth sets the limit to the top frequency response in analog transmission, while the pulse dispersion sets the limit to the maximum pulse rate in digital transmission, without regeneration for a given length of fiber.

In fibers made with silicate glasses, the mechanisms causing loss are absorption and scattering. The electronic and molecular absorption of the material attenuates the optical energy by converting it to mechanical vibrations known technically as "phonons." This mechanical energy is dissipated as heat. The dominant absorptions are due to the infrared phonon absorption edge, the ultraviolet (UV) electron absorption edge of the constituent elements, and other absorption bands of impurity ions incorporated within the material such as transition ions and OH^- ions. Scattering dissipates energy by causing the optical energy to be directed in different directions to the direction of propagation, thus resulting in a reduction of energy in the intended direction of propagation. This mechanism is due to the presence of scattering centers such as density changes and particulate inclusions or voids. The absorption losses due to transition ions can be reduced to lower than the scattering loss limit and hence are seldom of significance. An OH^- ion, however, usually causes absorption losses strongly at the wavelengths corresponding to the resonance frequency of OH^- ions and to a lesser extent at its harmonics and beat frequencies.

An illustrative spectral loss curve is as shown in Fig. 1-2 showing the scattering loss dominated regions <1.6 μm plus the OH^- absorption bands and the infrared (IR) absorption edge influencing loss at wavelengths of >1.6 μm. It is to be noted that other glass systems have different IR absorption edges and could be used in the >1.6-μm spectral range, where scatter loss is much lower.

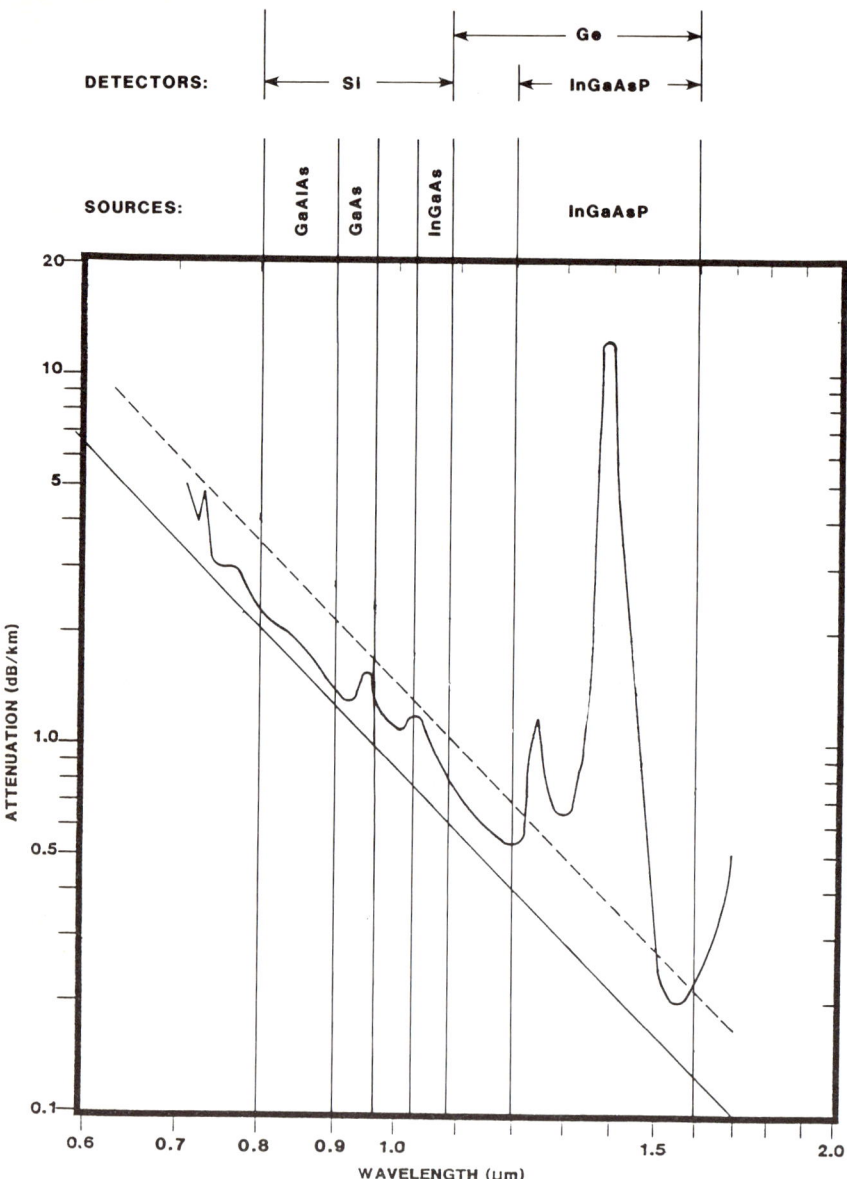

Fig. 1-2 Optical fiber attenuation versus wavelength.

The Fourier transform of the impulse response is the bandwidth response of the fiber. The dispersion of a length of fiber is governed by the difference in the time of arrival of the optical energy impulse launched at one end and received at the other end. The design of the fiber can substantially alter the dispersion. Dispersion is caused by the waveguide propagation characteristics, the spectral width of the light source, and the fiber design. Dispersion values of several tens of nanoseconds per kilometer down to a few picoseconds per kilometer are possible. The equivalent information bandwidths are tens of megahertz-kilometers to tens of gigahertz-kilometers, respectively. Since optical carrier frequency is in the region of 10^{14} to 10^{15} Hz, even a bandwidth of 10 GHz is a small fractional bandwidth and would not experience differential attenuation. Hence, for optical fiber systems, fiber offers constant attenuation for any operating bandwidth, in contrast with copper cables, where the attenuation of the cable increases as (bandwidth)$^{1/2}$. It is a factor which simplifies system design.

SPECIAL FEATURES OF OPTICAL FIBER WAVEGUIDES

Several features are unique to optical fiber waveguides. These can be utilized advantageously for special system applications. First, optical fiber waveguides are made with inorganic glasses. Typical softening point of high-silica glasses is at least 600°C. The refractive index change with temperature is about 0.0001/°C. The waveguide itself can operate easily over a temperature range greatly in excess of −55 to 125°C. In fact, it should work readily from −250 to 500°C. In practice, the temperature endurance of the coating material over the fiber determines the operational temperature limit. Thus the first special feature is its wide temperature working capability.

Optical fiber carries electromagnetic energy at optical wavelengths. Since glass is a dielectric and since fiber dimension is much smaller than the wavelength of electromagnetic waves in the radio-wave and microwave wavelengths, it does not pick up such radiation. It is immune to electromagnetic interferences. This is the second special feature, and it allows fiber to pass through regions of high electromagnetic fields and carry the information through without interference. Furthermore, the optical waveguide structure does not permit optical energy to be coupled from the side, thus ensuring no interference even from electromagnetic energy at optical wavelengths. (Note: Because of imperfections, there is a finite level of interference possible, and if the waveguides are placed side by side over long distances, coupling of energy from one fiber to another is possible.)

Since the radiation from the optical fiber is at optical wavelengths, it will not cause interference at radio frequencies. A severed fiber will not create a noisy environment to electronic equipment, nor will it pose a fire hazard. These are the third and fourth special features of optical fiber which can be used advantageously, for example, in aircraft installations and in systems with improved security against privacy intrusion.

SUMMARY OF FIBER SYSTEM CHARACTERISTICS

Physically, the fiber is small, flexible, strong, lightweight, and inexpensive; optically, it has a large bandwidth for information carrying and low loss and special properties which allow the transmission path to withstand extreme temperature range, resist electromagnetic radiation interference, and cause no electromagnetic interference or fire. These fiber characteristics distinguish fiber systems from others.

Chapter

Algorithm for Optical Fiber Link Design

Introduction

The basic optical fiber link system is presented at the component level in Fig. 2-1. The system is designed to permit input signals to be transmitted and received over a given distance with an acceptable signal quality. The criteria for acceptability are determined by requirements set for a particular application.

In a system design the choice of the light source determines the optical signal power available, and the choice of the detector and its

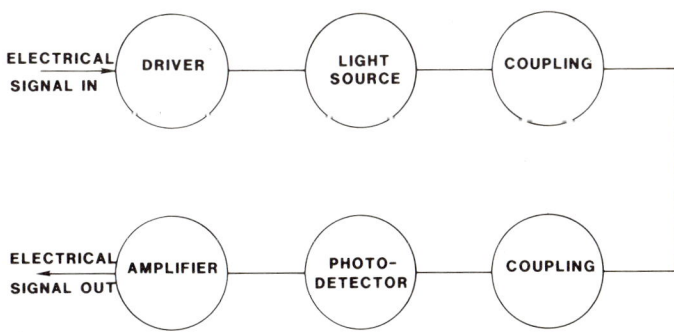

Fig. 2-1 A basic optical fiber link.

associated circuit determines the receiver sensitivity. The difference in the two signal levels is the allowable loss between the source and the detector, referred to as the *link margin*. The losses include the coupling losses at the junctions to and from the fiber transmission line and any intermediate joints, together with the loss along the length of the fiber. The allocation of different loss against the link margin is often referred to as the *preparation of a power budget*. Apart from power loss, the signal is also subject to distortions caused by nonlinearity of the light source and the delay variation along the optical fiber. The loss and the distortion influence the quality of the signal as it transmits through the system.

In order to discuss system design concepts without first fully explaining the desired features and critical characteristics of the system components, a relatively straightforward description is given for all the components by assigning a transfer function to each. These transfer functions are system design parameters which express primarily the input and output relationships. For example, the transfer function of a semiconductor light source can express the input drive current required for a given output light power. This and other transfer functions are now discussed.

LIGHT SOURCE

The input current-to-output power ratio is a system design parameter of a semiconductor light source. This defines the drive current needed to generate the optical output power. The choice of a particular light source for a particular system determines the electronic drive circuit needed to amplify the input signal to the required level to drive the light source and the optical signal power available from the source for coupling to the fiber.

With this parameter the designer cannot infer the deficiencies of the optical signal quality, optical source emission characteristics, power conversion efficiency, maximum speed of response, and a host of other relevant parameters. However, the designer has access to the two essential quantities to start the design: the drive current needed and the resulting signal power.

LIGHT SOURCE-TO-FIBER JUNCTION

A transfer function is required to define the power loss associated with the coupling of the signal power from the light source to the fiber transmission line. It is conveniently expressed in the ratio of optical power from the light source to the optical power launched into a given fiber.

The power from the light source is emitted over a solid angle ranging from omnidirectional and/or lambertian as in the case of a semiconductor light-emitting diode (LED), to a highly directional beam confined to a solid angle of a few degrees as in the case of a gas laser. The collection efficiency of a fiber depends on the core diameter and its acceptance angle or numerical aperture. Thus the source to fiber junction loss varies over a wide range, depending on the choice of the light source and fiber combination. In practice, the light source may be terminated in a fiber connector. In this case the light source output refers to the power available at the connector port.

For a semiconductor source and a 50-μm core fiber, the transfer loss can be as large as 20 dB, while for a laser to fiber, this loss can be as low as 3 dB. If light source power is specified as power at the connector port, the transfer loss for these devices is likely to be around 1 dB. It is to be noted that the unit "decibel" in optical fiber work generally refers to the optical power ratio 10 log (P_2/P_1), where P is the optical power.

FIBER

The transfer parameter of a fiber expresses the power launched into the fiber at one end to the power emerging from the other end. Since the fiber losses are insensitive to the bandwidth of the signal, the bandwidth aspect can be dealt with separately. As defined, this transfer parameter enables the power budget of the system to be readily calculated. Since fiber loss varies with wavelength, the transfer parameter for a fiber differs for each spectral region. A plot of transfer function versus wavelength is identical to the spectral loss curve as shown in Fig. 1-2.

The signal bandwidth of a fiber is dealt with in terms of pulse delay and intersymbol interference level if a digital signal is involved, or in terms of the bandwidth if an analog signal is involved.

A bandwidth analysis is an important and more involved part of the system design. A relatively inexact but useful first cut for digital signals can be done by expressing the fiber delay characteristics in nanoseconds per kilometer; that is, pulse width will increase by x ns over a distance of 1 km. Thus, to avoid overlap of successive pulses, the designer must ensure that the delay or pulse broadening be smaller than the time interval between the pulses after propagating along the entire length of the fiber in question. The delay is taken to be linearly additive. If a fiber has a delay of 1 ns/km, it will handle the transmission of a 100-Mb/s signal in return to zero-pulse format over ∽5 km.

FIBER-TO-FIBER JUNCTION

The loss involved in either joining the fiber by a permanent splice or the use of a demountable connector is a transfer parameter to be included.

Splices are used to make permanent joints in order to increase fiber cable length or to execute cable repairs. Splice losses are generally rather small. Connectors are used to make demountable connections. They are used to allow easy connection and disconnection between cables and equipment. This transfer parameter is conveniently expressed in decibels, with 0.2 dB for a splice and 1 dB for a demountable connection as examples.

FIBER-TO-DETECTOR JUNCTION

At the input to the receiver a fiber cable terminates at a photodetector. An additional optical power loss may be incurred if the light exiting from the fiber is not wholly collected by the photodetector. This loss is expressed as a transfer parameter. Usually this loss is very small, in the order of 0.1 dB, contributed principally by the losses incurred through reflections at the detector surface and at any intermediate glass seal over the detector.

PHOTODETECTOR

The reconversion of optical power into electrical power takes place at the photodetector. The photodetector performance can also be expressed as a transfer parameter which is the ratio of electrical current generated at the detector to optical power into the detector in amperes per watt. The spectral response, the speed of response, and the linearity of the photodetector are also relevant as in the case of the light source.

The photodetectors generally produce photocurrents which must be further amplified by suitably designed electronic current. The photodetector-amplifier combination is referred to as the *receiver*. The receiver design is aimed at achieving high sensitivity and low distortion.

Design Algorithm

A basic fiber link is to be designed as a system to provide a 5-km transmission path capable of handling a signal at 45 Mb/s in return-

Algorithm for Optical Fiber Link Design   19

to-zero (RZ) format; that is, adjacent pulses are separated by a zero-amplitude pulse. The design algorithm is as follows.

Select a light source, a detector, and a fiber with a speed of response and dispersion capable of handling the 45-Mb/s signal in RZ format. Obviously, the source and the detector must match spectrally and must be in the spectral region where a desired fiber loss is attainable.

Assuming the transfer function for the items selected are as shown in Table 2-1, a power budget is derived and is shown in Table 2-2.

If the receiver design requires a minimum of −42-dBm (decibels above 1 mW) peak signal for satisfactory recovery of the transmitted signal, the leftover margin is 7 dB. A margin of several decibels is needed to accommodate system parameter changes over the operating temperature range and over the life of the equipment.

This description of the algorithm of a system design highlights the need to know certain basic characteristics of the components to execute a basic design and alerts the reader to the fact that more knowledge of the performance of the components related to their bandwidth, distortion, noise, and loss characteristics must be considered in order to develop technically effective designs. Installation, environment, cost, and other operational conditions must also be taken into account at component selection level as well as at system design level. In practice a design is a multidimensional exercise. A thorough knowledge of component characteristics is needed to permit design trade-offs.

Table 2-1 **Typical Transfer Function Values**

LED light source	100 mW/A
LED source to fiber coupling loss	20 dB
Fiber loss	4 dB/km
Splice loss	0.5 decibel per splice
Fiber-to-detector coupling loss	0.1 dB
Detector	0.5 A/W

Table 2-2 **Power Budget**

Power from light source 100 mW/A with 100-mA drive current	+10 dBm
Light source-to-fiber coupling loss	20 dB
Fiber loss, 5 × 4 dB/km	20 dB
Splice loss assume 10 required at 0.5 dB/joint	5 dB
Fiber-to-detector coupler loss of 0.1 dB—neglected	—
Total loss	45 dB

Hence, power available at the detector is −35 dBm.

Chapter 3

Fiber

Introduction

Central to optical fiber systems is the optical fiber waveguide. It is a threadlike structure. In its simplest form it has a light-guiding region, referred to as the *core,* surrounded by a layer of material, a coaxial outer region, known as the *cladding.* The optical fiber is designed to work as a transmission line to conduct electromagnetic energy of a particular wavelength. The information-carrying capacity of the fiber depends on the fiber design, the fiber material properties, and the spectral width of the electromagnetic energy source.

The principle of operation of the fiber is explained rigorously by electromagnetic theory or, less accurately but with good pictorial clarity, in terms of geometric optics. Total internal reflection, which occurs when a light beam emerges from a denser to a rarer medium, is the basic mechanism involved in the transmission of light along the fiber.

Practical fibers are designed to fulfill different functions. The single-mode fiber has ultimate wide bandwidth, while graded-index fiber provides adequate information-carrying capability combined with relative ease of handling. Large-core step-index fiber is convenient when a maximum amount of light is to be collected from the light sources. The characteristics of these fibers depart from the ideal as a result of physical imperfections, such as material inhomogeneity and lack of dimensional precision.

In operation the fiber encounters different environments and experiences various forces. Under these conditions the basic fiber strength characteristics are found to be excellent, even though accelerated failure under stress known as *fatigue* can occur. The effect on the transmission properties through the introduction of bends is significant and must be taken into account in fiber cable designs.

These different aspects of fiber properties and the methods of evaluation are the topics of this chapter.

Physical Structure

An optical fiber is a long cylindrical structure, usually with a circular cross section. In its simplest form it has two coaxial regions. The inner region is the light-guiding core, while the cladding completes the light-guiding region in such a way that the outer surface can be handled with little disturbance to the propagation characteristics. Different types of fibers can be classified into two general categories. The fiber with a core of constant refractive index n_1 surrounded by a cladding of constant refractive index n_2 is referred to as a *step-index fiber waveguide*. For light guidance, $n_1 > n_2$ is required. The fiber with a core of gradually varying refractive index is referred to as a *graded-index fiber waveguide*. The profiling of the index gradient alters the information bandwidth capacity of the fiber.

In a practical fiber the core and the cladding may be conveniently surrounded by a second coaxial layer which serves as a supporting structure, to increase physical robustness. The overall size of the fiber and the overall-diameter to core-diameter ratio control the sensitivity of the fiber propagation characteristics to bending.

Basic Theory

Wave propagation along a fiber waveguide can be formally set up as a boundary value problem in electromagnetic theory.[1] The solution of a dielectric rod in an infinite medium of lower refractive index was solved in 1910 by Hondros and DeBye,[2] thus establishing the characteristics of electromagnetic wave propagation along a long cylinder. To gain insight into the nature of propagation within a fiber waveguide, it is appropriate to start by understanding the phenomenon of reflection and refraction at a dielectric interface.[3]

REFLECTION AND REFRACTION AT A DIELECTRIC INTERFACE

When a plane wave front is incident at a boundary of two dielectrics with different refractive indices n_1 and n_2, the incident wave is reflected and refracted at the boundary as shown in Fig. 3-1.

A rigorous solution can be obtained by solving the Maxwell equations applied to this particular boundary condition. It is found that the angle

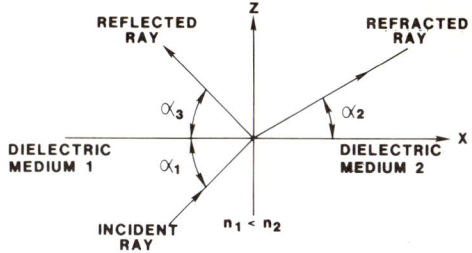

Fig. 3-1 Light rays at a dielectric interface $x = 0$.

of reflection α_3 is always equal to the angle of incidence α_1; that is, $\alpha_1 = \alpha_3$. This is the well-known law of reflection in optics.

The angle of refraction α_2 is found to satisfy the relationship

$$\frac{\sin \alpha_1}{\sin \alpha_2} = \frac{n_2}{n_1}$$

This is known as Snell's law.

For $n_1 > n_2$, $\sin \alpha_2$ approaches 1 as α_1 increases to $\alpha_c = \sin^{-1}(n_2/n_1)$. This angle is referred to as the *critical angle*. When α is infinitesimally greater than α_c, no refracted wave can exist. This effect is known as the *total internal reflection*. It can be shown that the wave solution is still valid when $\alpha > \alpha_c$. This allows a more detailed look at the refracted wave. It transpires that the refracted wave is no longer propagating but is decaying rapidly beyond the boundary. The greater α exceeds α_c, the more rapid the rate of decay. The critical angle allows the light to be bounded within the denser medium and permits lossless

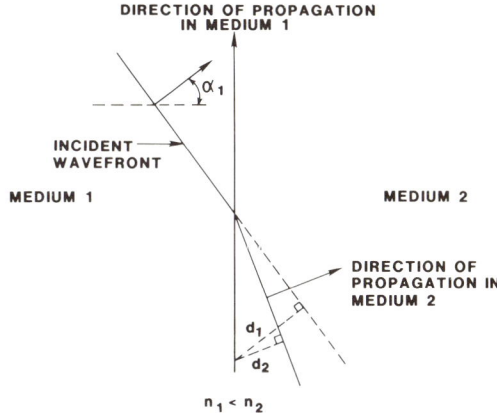

Fig. 3-2 Wave front crossing a dielectric boundary (Snell's law).

propagation, referred to as *bounded* or *nonradiative* propagation, within a fiber.

It should be noted that the velocities of propagation in media 1 and 2 are c/n_1 and c/n_2, respectively, where c is the velocity of light in vacuum. The refracted wave front must change direction to account for the velocity difference.

The relative distance covered by the wave front in medium 1 to medium 2 is

$$\frac{\sin \alpha_1}{\sin \alpha_2} = \frac{d_1}{d_2} = \frac{n_2}{n_1}$$

which is Snell's law. This is illustrated in Fig. 3-2.

TRANSMISSION AND REFLECTION COEFFICIENTS

Transmission and reflection coefficients are the ratios of the power transmitted and reflected relative to the incidence power, respectively. These coefficients can be computed by considering the power flow. Whenever light leaves or enters fiber waveguides and passes between different waveguide structures, these coefficients allow an estimation of the reflected and transmitted energy.

When the electric field vector is parallel to the boundary plane, the transmission and reflection coefficients vary with the angle of incidence. The transmission coefficient is a maximum for normal incidence and is given by a simple expression:

$$T_E = \frac{4n_1 n_2}{(n_1 + n_2)^2}$$

The reflection coefficient is a minimum for normal incidence and is given by another simple expression:

$$R_E = \frac{(n_1 - n_2)^2}{(n_1 + n_2)^2}$$

For the case of a glass-air interface with $n_1 = 1$ and $n_2 = 1.5$, R_E is approximately 4 percent.

These are valid for $n_1 > n_2$ or $n_1 < n_2$. However, if $n_1 > n_2$,

$$T_E = 0 \quad \text{at} \quad \alpha_1 \geq \alpha_c$$

and $R_E = 1$

while for $n_1 < n_2$

$$T_E = 0$$

and $R_E = 1$ at $\alpha_1 = 90°$

When the electric ventor is normal to the plane of incidence, the power transmission and reflection coefficients also exhibit angular dependence. For normal incidence, these reduce to the same formula as in the first case. The condition for total internal reflection is the same. However, R_H vanishes when

$$\sin \alpha_1 = \frac{n_2}{(n_1^1 + n_2^2)^{1/2}}$$

This angle is known as the *Brewster angle*.

PLANE WAVE IN A STRATIFIED MEDIUM

A stratified medium consists of two or more boundaries. The laws of reflection and refraction apply at each boundary (Fig. 3-3).

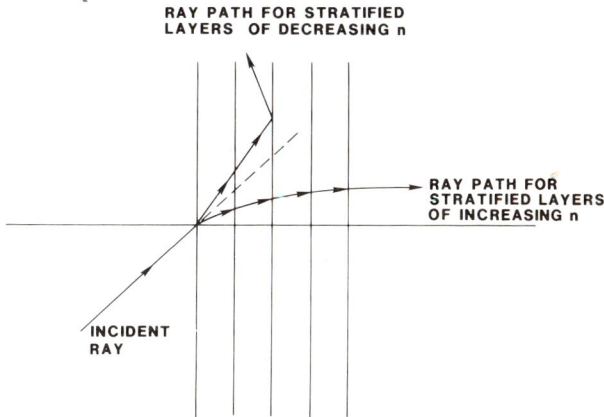

Fig. 3-3 Ray propagation in a stratified medium.

If the incident plane wave is propagating within a medium toward a stratified region of successively lower index, the angle of incidence at each succeeding boundary will increase until it reaches or exceeds the critical angle, when total internal reflection will take place. This is an important phenomenon encountered in the graded-index optical fiber waveguide. It is commonly met in nature on hot and humid days when the warm road surface can cause the heavy air to form a graded-density region immediately above the road surface. The air will have an increasing density away from the road surface such that a person viewing the road from the driver's seat of a car can see the mirage effect when the road surface appears to be wet and mirrorlike.

Chapter Three

OPTICAL FIBER WAVEGUIDES

Electromagnetic wave propagation within an optical fiber waveguide can be approximately described in terms of the plane wave propagation.[4] The cylindrical geometry of the optical fiber makes the plane-wave analogy difficult to envisage. Nevertheless, the analogy is most pertinent. It is especially useful to provide a pictorial impression of the wave propagation when the plane waves are represented by light rays. In a cylindrical fiber waveguide the meridional rays can most readily be visualized. Take a longitudinal cross section of the cylinder as shown in Fig. 3-4. A meridional ray meeting the boundary at an angle greater than α_c will mostly radiate as a refracted ray while the reflected portion will be very small. Another ray meeting the boundary at an angle less than the α_c will be totally internally reflected. Such rays will propagate down the fiber, bouncing losslessly at the boundary. Recalling that the region outside the fiber boundary has an evanescent field at the position where the total internal reflection takes place, the immediate vicinity of the fiber boundary must not be disturbed. Any perturbation will have a profound effect on the mechanism of total internal reflection.

The propagation of skew rays is less straightforward, since these rays strike the boundary at an angle to a local tangent plane which is not parallel to the fiber axis. Thus it is difficult to determine whether a ray is totally internally reflected at the boundary or partially reflected and refracted. The geometry of the skew ray is as shown in Fig. 3-5.

The skew ray incident at angle θ with respect to (wrt) the Z direction and ϕ wrt the local tangent plane at P_1 is obtained by projecting the skew ray to the cross-sectional plane and measuring the angle between the projected line and the tangent at P.

The skew ray which makes an angle $\theta > \theta_c$ with the Z axis can incident at an angle $\phi < \theta_c$ to the tangent plane at point P. The rays which make angle $\theta < \theta_c$ are bounded while those with $\theta > \theta_c$ but $\phi \leq \theta_c$ have been shown to be partially refracted. These are sometimes called "leaky" rays, since they are slowly radiative. These rays can persist over a significant distance after launching. (Note: θ_c is $90 - \alpha_c$.)

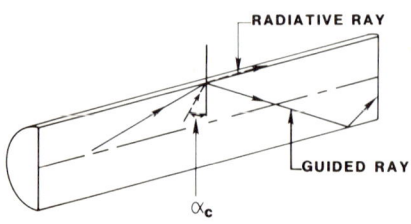

Fig. 3-4 Ray propagation within a cylindrical fiber.

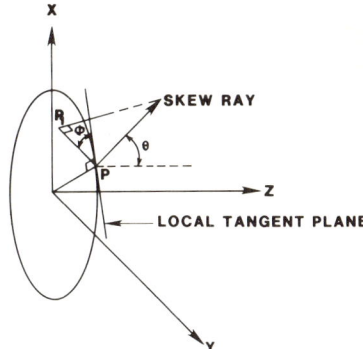

Fig. 3-5 Geometry of a skew ray.

The ray representation does not provide an exact description of the wave propagation characteristics but offers a readily visualized description and can give clues to explain transient and perturbation phenomena. The exact steady-state description is obtained by formulating the optical fiber waveguide as a boundary-value problem of electromagnetic theory.

ELECTROMAGNETIC FORMULATION

The simplest fiber structure which can be solved exactly is an infinite uniform cylinder of refractive index n_1 in an infinite medium of refractive index n_2. Obviously, this is not a practically usable structure.

By solving the Maxwell equations for the given boundary condition, the exact characteristic equation, or the eigenfunction, can be determined. The solutions, or eigenvalues, represent the propagation modes of this structure. The characteristic equation is

$$\left(\frac{\epsilon_1}{\epsilon_2}\frac{a\gamma^2}{k}\frac{J_\nu^1(ka)}{J_\nu(ka)} + i\gamma a\frac{H_\nu^1(i\gamma a)}{H_\nu(i\gamma a)}\right)\left(\frac{a\gamma^2}{k}\frac{J_\nu^1(ka)}{J_\nu(ka)} + i\gamma a\frac{H_\nu^1(ka)}{H_\vartheta(ka)}\right)$$

$$= \left[\nu\left(\frac{\epsilon_1}{\epsilon_2} - 1\right)\frac{\beta k_2}{k^2}\right]$$

where $\epsilon = \epsilon_0 n^2$ relative permittivity
$k = (n^2 k_0^2 - \beta^2)^{1/2}$ at cutoff $k = nk_0$, radial wave number
$k_0 = 2\pi/\lambda_0$, free-space propagation coefficient
$\gamma = \alpha + j\beta$, longitudinal propagation coefficient
a = fiber radius
ν = order of Bessel function J and Hankel function H

A solution is obtained when a β value satisfying the equation is found corresponding to a given set of ϵ, k_0, a, and ν. Each solution is referred to as a *mode of the fiber waveguide*. For each mode, the electromagnetic field assumes a unique distribution with radial and circumferential variations. The modes are designated by ν and μ, where ν is the order of the Bessel and Hankel functions and μ is the nth root of the Bessel function $J_\nu(ka) = 0$. A mode is at cutoff when $\beta = 0$. The cutoff of the lowest-order $\nu = 1$ mode is given by $J_0(ka) = 0$ or $ka = 2.4$, the first root of J_0. The single mode operation is defined by the relationship $ka = (2\pi/\lambda_0)a\,(n_1^2 - n_2^2)^{1/2} = 2.4$ for a fiber core of refractive index n_1 in an infinite cladding of refractive index n_2. The equation is complex and solutions can be obtained only numerically. For a structure with lossless material, the propagation coefficients of bounded or nonradiative modes for a number of lower-order modes have been determined.[5] When $n_1 < n_2$, no bounded mode exists, as expected.

In a practically useful situation the value of n_1 and n_2 for a fiber is relatively close such that Δ is small in the equation $(n_1 - n_2)/n_2 = \Delta$, allowing approximations such as $n_1 = n_2 = n$ and $\beta = nk_0$ to be used.[6] This proves to be useful in simplifying the solutions. Introducing this approximation, the characteristic equation is reduced to

$$\frac{kJ_{\nu-1}(ka)}{J_\nu(ka)} = \frac{j\gamma\,H^{(1)}_{\nu-1}(j\gamma a)}{H^{(1)}_\nu(j\gamma a)}$$

The closed-form solution can now be constructed for both near to cutoff and far from cutoff conditions.

If the field near the interface is examined, it can be seen that for bounded modes, the field decays exponentially in the radial direction. This means that some distance away from the interface the field is virtually nonexistent. Thus the outer medium can be terminated without substantially changing the field distribution. It also means that a coaxial structure with a core surrounded by a cladding layer can be constructed for easy handling. The propagation characteristics remain substantially the same as the cylindrical fiber in an infinite medium. This coaxial structure is the basic practical fiber.

The existence of the outer boundary is significant to radiative modes, since the outer boundary traps the radiated energy escaping at angles below the critical angle of the cladding and the outside medium. In fact, the coaxial two-layer fiber can be solved as a composite waveguide. The radiative behavior can then be more precisely defined. However, since the cladding dimension is relatively large, the discrete spectrum of allowed modes values are sufficiently dense to be considered as a continuum, approximating the true radiative modes.[7] This situation must be further refined when dealing with waveguides with composite cladding and supporting structures.

A summary of the modal description of an idealized fiber waveguide is given here. The discrete bound modes consist of the hybrid $HE_{v\mu}$ and $EH_{v\mu}$ modes each with two orthogonal directions of polarization. Except for the HE_{11} mode, each mode has a cutoff; that is, as wavelength increases, a mode will change from a bounded to a radiative mode at a particular wavelength. The HE_{11} mode remains bounded as $\lambda \to \infty$. At cutoff the mode radiates along the direction of the fiber axis and propagates at a velocity c/n_2. Far from cutoff, modes propagate with velocities spreading from c/n_1 to c/n_2. The number of modes is a function of the fiber core radius and operating wavelength. Using a convenient normalized frequency parameter

$$V = \frac{2\pi a}{\lambda} (n_1^2 - n_2^2)^{1/2} \sim n_1 \sqrt{2\Delta}\, ka$$

the total number of modes N, in a waveguide with core index n_1 and cladding index n_2, is approximately $N = V^2/2$.

The characteristic function for the exact solution is too complicated for analytical solutions. Little can be deduced from it to provide general descriptions of modal characteristics.

The approximate eigenfunction still requires numerical solution. However, asymptotic expressions can be derived to provide useful information such as dispersion, mode, cutoffs, and so on. The most pertinent limitations and validity of the simplified solution have been concisely summarized by Marcuse:[8]

The exact eigenvalue equation has twice as many solutions as the simple equation. The exact field solutions of the round optical fiber are classified as $HE_{v\mu}$ or $EH_{v\mu}$ modes. The propagation constants of $HE_{v+1,\mu}$ and $EH_{v-1,\mu}$ modes are almost identical. They become exactly the same in the limit $n_1 \to n_2$. We are thus faced with a case of near degeneracy. Comparison of the simplified mode solutions with the exact modes shows that the simplified modes are actually a superposition of $HE_{v+1,\mu}$ modes. The near degeneracy of the exact theory has thus become a definite degeneracy, and the two types of modes have merged into one. However, the total number of modes is the same in both theories, because we now have a fourfold degeneracy since both polarizations and both choices of sine or cosine functions lead to the same eigenvalue equation.

The dispersion curves, representing the propagation constants as functions of frequency, are very nearly the same for the simplified and exact modes in case of weakly guiding fibers. Owing to the near degeneracy of the HE and EH modes, their dispersion curves are almost indistinguishable. The simplified description is thus able to reproduce the dispersion characteristics of the modes. This enables us to study the problem of pulse distortion with the use of the simplified eigenvalue equation. Problems of mode conversion and radiation losses can also be studied with the help of the simplified modes. Instead of determining how each *HE* or *EH* mode couples to other modes we now find how the superpositions of $HE_{\nu+1,\mu}$ modes and $EH_{\nu-1,\mu}$ modes couple to each other and to radiation. For purposes of determining the power transfer between groups of guided modes and for the study of radiation losses we can gain all the information that is required.

However, in spite of the obvious advantages of the simplified theory, it is prudent to keep in mind that the simplified modes do not represent true modes in the usual sense of the word. Even though we cannot determine this fact from the approximate analysis, comparison with the exact theory teaches us that the simplified modes must decompose as they travel along the waveguide. Because they are actually superpositions of $HE_{\nu+1,\mu}$ and $EH_{\nu-1,\mu}$ modes that travel with slightly different velocities, the simplified modes change their shape as they travel along the guide. This feature of the simplified modes becomes clear when we realize that the field shape of the superposition of two modes depends on their relative phase relationships. Because of their different phase velocities the relative phases of the $HE_{\nu+1,\mu}$ and $EH_{\nu-1,\mu}$ modes keep changing as a function of z, so that the superposition fields also change their shape. Only after a distance corresponding to one beat wavelength does the original relationship, and, therefore, the field shape, restore itself.

The approximate modes are labeled as $LP_{\nu\mu}$ modes: LP_{01} corresponds to HE_{11}, and $LP_{\nu\mu}$ corresponds to a superposition of $HE_{\nu+1,\mu}$ and $EH_{\nu-1,\mu}$ modes.

The propagation constant β for each mode can be deduced from computed value of a normalized propagation constant b as a function of V. This is shown in Fig. 3-6. The relevant approximate equation is

$$\beta = (b\Delta + 1)n_2 k$$

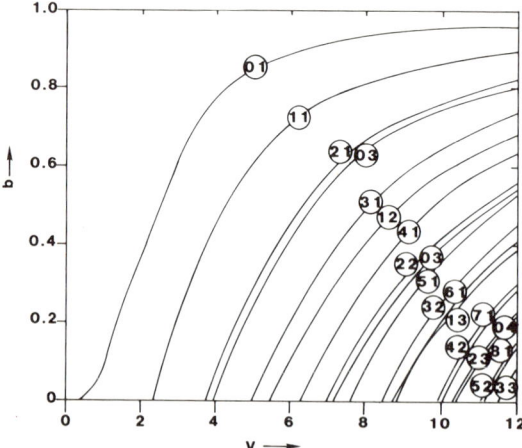

Fig. 3-6 Propagation constant of *LP* modes.

Fiber 31

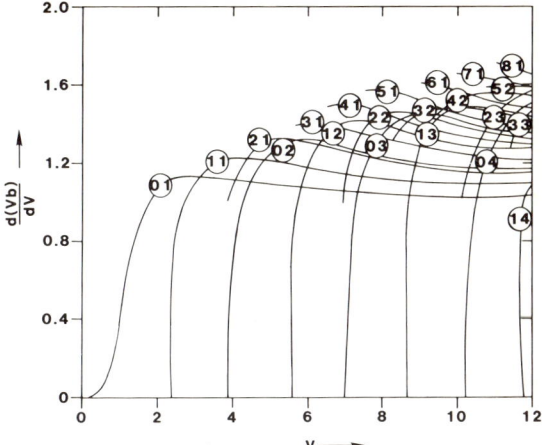

Fig. 3-7 Waveguide group delay of *LP* modes.

The group velocity $V_g = dw/d\beta$ or the group delay $\tau_g = (d\beta/dw)$ through a length of waveguide of length L is deduced as follows:

$$\tau_g = L\frac{d\beta}{dw}$$

$$= L\frac{d}{dw}(b\,\Delta n_2 k + n_2 k)$$

and after manipulation

$$= \frac{L}{c}[n_2\,\Delta d(Vb)/dV + d(n_2 k)/dk]$$

The first term is the waveguide delay, and the second term is the delay caused by material dispersion. The term $d(Vb)/dV$ is the normalized group delay. Computed values for different modes are as shown in Fig. 3-7.

SUMMARY OF OPTICAL FIBER WAVEGUIDE THEORETICAL DESCRIPTIONS

The modal description provides an exact means of depicting wave propagation within a fiber in accordance with Maxwell's electromagnetic wave theory.

Because of the complexity of the exact solution, approximation has been introduced. Using a practically relevant approximation of

$n_1 \sim n_2$, a much simplified eigenfunction has been obtained. This approximation has definable limits and has provided important information such as group velocities which clarify the transmission characteristics of each mode. It also provides an effective basis for constructing theoretical solutions, such as for problems of waveguide discontinuities.

Certain modal fields can be generated by considering the transverse resonance condition of conical rays. The modes have characteristic angles and can be linked perfectly with the ray description in terms of meridional rays.

In general, modal fields cannot be linked on a one-to-one basis to ray description. The ray representation is an approximation without properly defined limits. It can provide quantitative description, but more often it serves to give a pictorial but effective subjective description.

Comparison of Waveguide Types

SINGLE-MODE WAVEGUIDE

When a fiber waveguide can support only the HE_{11} mode, it is referred to as a *single-mode waveguide*. In a step-index structure this occurs when the waveguide is operating at $V < 2.4$ (see cutoff definition). The propagation characteristics are completely determined by the propagation characteristics of the HE_{11} mode for a single-frequency optical wave. The polarization of the mode is, however, unconstrained. This could introduce polarization-dependent propagation characteristic variations through waveguide geometrical imperfections and material birefringence. Otherwise, the single-mode waveguide offers the highest information-carrying capacity in a predictable manner. A bandwidth of 50 GHz·km is achievable at the designed wavelength.

In practice, a single-mode waveguide structure may have significant imperfections, in the form of core ellipticity and particularly strain-induced birefringence, which could impose limits to the information-carrying capacity. Random discontinuities in the form of material inhomogeneities and particulate or void inclusions could cause conversion of the guided energy into the radiative modes, resulting in an increased loss. Some of these radiative modes could become trapped within the finite cladding and propagate as cladding modes. At a subsequent discontinuity they could reconvert at the wrong phase into the core mode, causing delay distortion. For these reasons practical single-mode waveguides must be carefully designed to minimize these effects. For example, the following design criteria could improve the bandwidth of a single-mode fiber: a tight control on core diameter and ellipticity, matched core and cladding material thermal expansion coefficients, and a cladding covered by a lossy layer of material with equal refractive index.

For a high-silica glass fiber operating at $\lambda = 1$ μm, with a core refractive index of 1.47 and a cladding index of 1.458, the maximum core diameter for single-mode working is defined by the cutoff relationship:

$$\frac{2\pi a}{\lambda}(n_1^2 - n_2^2)^{1/2} = 2.4$$

From this relationship the core radius a for this particular fiber is 2.04 μm. A practical single-mode fiber may have a structure as shown in Fig. 3-8.

The limitation on the choice of cladding thickness is governed by the need to ensure that the field at the cladding boundary is negligible, so that the fiber can be handled without affecting the propagation characteristics or can be coated by a lossy outer coating without being similarly affected. The required thickness can be precisely defined. In practice, over 10 times the operating wavelength has been found to be adequate when operating at V close to 2.4. However, because of the requirement to control loss due to bending, the cladding-to-core ratio is usually made much larger. The cladding which must be made with material of low loss is often coated with a supporting structure to achieve a rather large fiber outside diameter to core ratio.[9]

In high-silica fibers the waveguide and material dispersions are often of opposite signs. This fact can be used conveniently to achieve a single-mode fiber of extremely large bandwidth over a spectral region for a source with a known spectral width.[10]

The bandwidth of the single-mode fiber is controlled by the dV_g/dk characteristics of the HE_{11} mode and the material dispersion. It is a function of the operating wavelength and V value.

A single-mode waveguide, with its large and fully definable band-

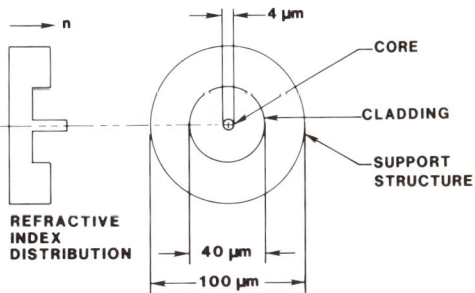

Fig. 3-8 Cross section of a single-mode fiber.

width characteristics, is an obvious candidate for long-distance, high-capacity transmission applications.

STEP-INDEX FIBER

A fiber waveguide with a core of uniform refractive index but operating with $V > 2.4$—that is, sufficiently large to support a number of modes—is referred to as a *multimode step-index fiber*. The propagation characteristics are governed by the various modes present. As the number of modes increases, the modal description tends to confuse rather than clarify. On the other hand, geometric ray description demonstrates clearly the principal characteristics.

It can be seen that an increase in core size and numerical aperture increases the amount of power which can be launched into the fiber from a lambertian light source with emission area greater than the fiber core. It can also be seen readily that the dispersion is controlled principally by the differential path lengths of the central ray (shortest path) to that of the ray propagating at an angle close to the critical angle. Furthermore, the waveguide performance is influenced by the distribution of power in the modes along the waveguide.

A typical multimode step-index fiber made with high-silica glasses may have a 100-μm core and a 140-μm outside diameter. The refractive index difference of the core and cladding are chosen to give a high numerical aperture of about 0.3. A typical structure is as shown in Fig. 3-9.

The choice of cladding thickness is almost arbitrary, since the perturbation on propagation is strongest only on the high-order mode nearest to cutoff. In a practical structure the cladding region sometimes serves as a barrier to prevent an impurity in the support tube from migrating or diffusing into the core during fabrication. It is also possible to make

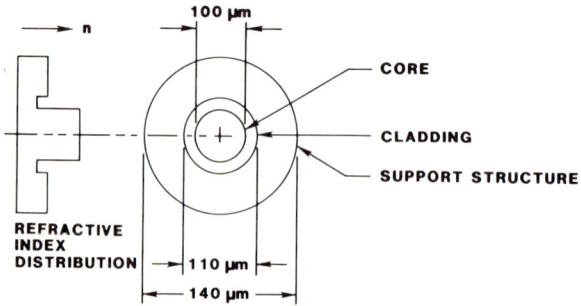

Fig. 3-9 Cross section of a multimode step-index fiber.

the cladding region with a refractive index lower than that in the support tube. This results in an increase in the numerical aperture.

The bandwidth and loss characteristics of the multimode step-index fiber are length-dependent, since waveguide imperfections and externally introduced bending cause mode conversion and redistribution.[11] The higher-order modes are more easily converted to radiative modes. In very long fibers with statistically random imperfections the mode distribution attains an equilibrium state. The bandwidth dependence on distance approaches $L^{1/2}$. At a particular position along the fiber the mode distribution could vary randomly. The minimum bandwidth of a multimode fiber is dependent only on numerical aperture (NA) of the fiber but not on the core diameter. A typical fiber of 0.25 NA has a usable bandwidth of around 20 MHz. It is suited for short-haul applications.

GRADED-INDEX FIBER

By varying the radial refractive-index profile,[12] a fiber with wider bandwidth capability is created. These fibers are referred to as *graded-index fibers*. Fiber profiles resulting in substantial bandwidth increase can be conveniently represented by a power law profile:

$$n_r = n_1 (1 - ar^n)$$

The characteristics of multimode graded-index fiber can be described to a sufficient degree of accuracy by using a ray description.

A typical multimode graded-index fiber made with high-silica glasses may have a 50-μm core and a 125-μm outside diameter with a maximum NA of around 0.2. A structure is as shown in Fig. 3-10.

The limitation on the choice of cladding thickness is as discussed

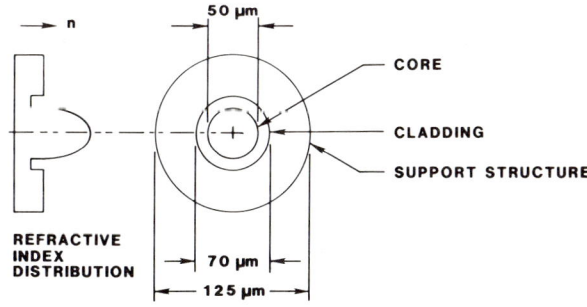

Fig. 3-10 Cross section of a multimode graded-index fiber.

36　Chapter Three

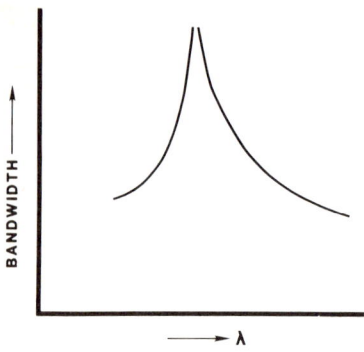

Fig. 3-11 Bandwidth as a function of operating wavelength of a graded-index fiber.

in the step-index case. However, the cladding refractive index can affect the bandwidth characteristics if there is an abrupt index change in the core-to-cladding boundary.

The bandwidth and loss characteristics of the graded-index fiber are dependent on the excitation condition, the fiber length, the light source line width, and the emission center wavelength for a given index profile.[13] The dependencies on excitation condition and fiber length are the results of waveguide imperfection and bending as discussed in the step-index case. The dependence on the line width and the center wavelength is due to the variation of material refractive index with wavelength.

For a given profile, the bandwidth versus wavelength is typically as shown in Fig. 3-11 but can be modified, by suitable choice of materials with appropriate material dispersions, to broaden the cusp of the bandwidth-versus-wavelength curve.[14]

The maximum bandwidth and the sharpness of the curve are dependent on the material used, the source line width launching condition and fiber length and dimensional tolerances. The fabrication technique imposes a bandwidth limit of about 1 GHz·km. It is a high-quality fiber suited for large-bandwidth, medium-haul applications.

PLASTIC-CLAD FIBER

By replacing the cladding and support structure with a plastic coating of refractive index lower than that of the core, a plastic-clad fiber is achieved. The core could be of homogeneous composition or with radially graded index. The replacement of the glass cladding with a plastic offers possible saving in cost, since the plastic coating could be cheaper than the glass support.

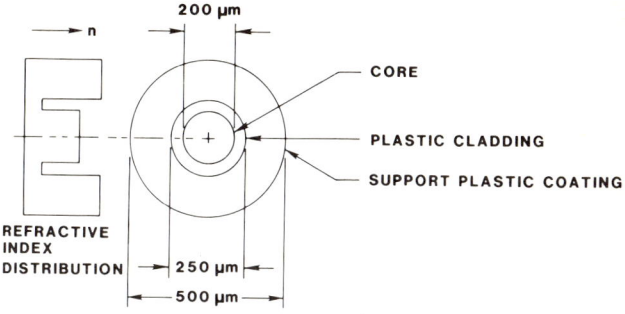

Fig. 3-12 Cross section of a plastic-coated silica fiber.

The plastic-clad fiber has a number of limitations. The plastic loss sets the lower fiber loss limit; the long-term durability of the plastic sets the fiber life, especially in a humid environment; the refractive-index change with temperature sets the lower temperature limit; and the softening point of the plastic sets the upper temperature limit. A typical structure is shown in Fig. 3-12.

The plastic clad fiber is suitable for special applications. For example, a silicone-coated pure silica fiber has good resistance to radiation-induced loss increase and a silicone-coated pure silica fiber with large core diameter of several hundred micrometers serves as a convenient large-core high-NA fiber. The bandwidth available is of the order of 20 MHz·km. Because of the lossy cladding, this type of fiber has a loss of several decibels per kilometer.

Fiber Imperfections

In a practical fiber the ideal waveguide structure cannot be obtained over an indefinite length. Imperfections are present on a statistical basis. These imperfections are the basic material and fabrication process variations, resulting in the formation of scattering centers and dimensional variations, which are distributed statistically throughout the fiber length. They give rise to additional losses and modify bandwidth characteristics. Even with all the imperfections, fiber loss and bandwidth approaching the theoretical limits are achievable in practice.

BASIC MATERIAL PROPERTIES

For amorphous materials such as organic polymers and inorganic glasses, random fluctuation of material composition is present. The glassy-state solid retains some of the fundamental behavior of the liquid state.[15] The localized material density fluctuations give rise to a scattering loss which is given by

$$36 \times 10^3 \frac{(n-1)^2}{\lambda^4} KT_{\beta c} \text{ dB/m}$$

where
- n = refractive index
- λ = operating wavelength in meters
- K = Boltzmann constant
- $T_{\beta c}$ = fictive temperature; that is, temperature at which glass viscosity has increased to a value when the liquid behavior is frozen

This loss is about 1 dB/km at 1-μm wavelength, and as indicated by the scattering loss formula, it will change with the composition of the glass which affects $T_{\beta c}$ and n. The fluctuations are microscopic since the scattering loss obeys the Rayleigh scattering law of λ^{-4}.

Occasionally, particulate or void inclusions could give rise to significant losses due to point scattering of large amounts of energy.

For a given glass composition, the scattering loss could increase as a result of composition fluctuation. Depending on the glass-forming process, the glass-forming elements may not be present in equal qualities throughout the bulk of the glass thus formed. Furthermore, the process of refining the glass varies in different fiber-making processes. For the chemical vapor deposition, the refining proceeds locally and does not correct the lack of homogeneity along the axial direction.

Material fluctuations generally are too small to cause a reduction in the bandwidth of the waveguide, although longitudinal inhomogeneity may have an effect on the fiber bandwidth.

DIMENSIONAL TOLERANCE

The fabrication process gives rise to dimensional fluctuations, expressible as diameter variations, ellipticity, eccentricity, and birefringence. These imperfections cause mode conversion and result in loss increase and modification to the bandwidth performance.

Diameter Variations

Changes in the fiber core and cladding dimensions usually occur simultaneously. They can be introduced through the dimensional variations of the starting rod, the lack of mechanical accuracy of the apparatus for fiber making, or variation in the heating conditions. For a fiber made by the internal chemical vapor deposition method, the starting substrate tube may have thickness variations and curvatures which resulted in the finished preform to have dimensional variations. At the fiber-drawing stage the heating condition may not be constant or the mechanical drives sufficiently precise. The net result would be a fluctuation of fiber diameter.

The effects of diameter fluctuation can be estimated by first understanding the effect of a sinusoidal perturbation. Coupled-mode theory can be applied to the case of sinusoidal variation to estimate the mode coupling between guided modes and between guided and radiative modes. The condition for perfect coupling between two modes is when the inverse of the difference of the phase velocity corresponds to the perturbation spatial frequency. Thus, for a more complex perturbation, the effect, in principle, can be estimated by first calculating the coefficients of all the sinusoidal components. However, the calculation based on coupled mode theory is extremely cumbersome, and meaningful results are difficult to obtain. It can be simplified by formulating the coupled-mode equations into an equivalent set of coupled-power equations.[16] These are derived by assuming the coupling to be weak, and deal only with the average powers carried by the guided modes.

The diameter fluctuations can be seen to cause an excess loss due to the coupling of power to radiative modes.[17] It also causes intermode coupling in such a way that the effective bandwidth of the waveguide is increased.[18] In principle, it is possible to increase the bandwidth with a controlled increase in loss by deliberate introduction of small-diameter fluctuations. However, it is likely to cause difficulties in the fabrication process control.

If a steady heat source and diameter monitoring with a feedback control are used, 2 percent tolerance is readily achievable. The fluctuation periodicity is sufficiently long that the resulting fiber exceeds loss is virtually zero. Under these conditions, the effect of mode conversion is also minimal.

Ellipticity

The fiber ellipticity is introduced during the fiber preform-making process but seldom during the fiber-drawing stage. The effect of ellipticity is to cause a polarization dependence. For the single-mode fiber, the elliptical fiber can support two orthogonal modes, one along the minor and the other the major axis of the waveguide.[19] These modes have different group velocities and hence would cause the effective bandwidth of the fiber to decrease. In the case of a multimode waveguide, small ellipticity is not critical. The set of elliptical modes behaves more or less similarly to the set of circular modes.

A deliberate elliptical single-mode fiber could be used to achieve a degree of control on polarization. It is difficult, however, to prevent mode coupling unless the fiber has high NA and strong ellipticity.[20]

Eccentricity

The fiber core eccentricity is introduced again during the fiber preform-making process but seldom during the fiber-drawing stage. Since the fiber cladding is usually sufficiently thick to allow the core to be somewhat eccentric without wandering to a region with no cladding material, the core eccentricity does not cause propagation characteristics to vary.

The major problem arises when two fibers are to be joined together. If the alignment is referenced off the outside diameter, fiber core eccentricity will cause jointing loss to increase.

Birefringence

The material inhomogeneity, and more particularly the difference in the thermal expansion coefficients of the core and cladding material, can give rise to local birefringence within the fiber. The effect of birefringence is to cause polarization fluctuations and, in the case of the single-mode fiber, bandwidth narrowing due to the different group velocity of the modes associated with the two polarization directions. The deliberate introduction of birefringence in a single-mode fiber can again be used to define the polarization of transmission within such a fiber.[21] In fact, the birefringence can be as effective as the ellipticity for controlling the polarization.

GENERAL EFFECT OF IMPERFECTIONS

The imperfections cause mode coupling and generally improve somewhat the bandwidth characteristics of the multimode waveguide and increase the waveguide losses. If a fiber waveguide is coated with a protective plastic with an index lower than that of the cladding or the support structure, the radiated power can propagate as a cladding mode. The power in the cladding mode may couple back into guided modes at discontinuities along the fiber. The effect is to cause a reduction of the effective bandwidth of the waveguide.

EXTERNALLY INDUCED IMPERFECTIONS

Besides structural imperfections, a fiber waveguide in use is subjected to externally applied forces which can result in twisting and bending.

Generally the amount of twist applied over a length of fiber seldom exceeds one turn per several centimeters and is more like one turn per several meters. The effect of twisting on the propagation characteristics is negligible except when a single waveguide has an elliptical cross section or a dominant polarization axis. In this case the twist will affect the output polarization direction. Bending, on the other hand, causes mode conversion. The sharper and more periodic is the bending along the fiber, the more pronounced is the mode conversion effect. A bending radius above 1 cm causes negligible propagation change. Hence the bending induced loss, which turns out to be the dominant effect, is referred to as *microbending loss*.

Microbending Loss

It can be shown theoretically that a curved fiber cannot support a set of nonradiative modes.[22] These modes are also different from those modes of the straight waveguide because they are radiative. Thus as a fiber curves there will be continuous losses as well as mode conversion at the transition between the straight and curved configurations. This results in mode redistribution and conversion into radiative modes.

In an optical waveguide, sharp curvatures involving local axial displacements of a few micrometers and spatial wavelengths of a few millimeters are termed *microbends*. Such bends may result from waveguide coating, cabling, packaging, or installation. Microbending can cause radiative propagation losses.

This way of looking at the curvature-induced loss indicates that random distribution of bends would induce mode redistribution and increase loss. The bending transition loss is a very steep function; the sharper the bend, the more pronounced the effect. Analysis has been carried out to estimate the sensitivity to externally applied forces on a given fiber structure.[23,24] Obviously, the stiffness of the fiber plays a role. Numerical aperture exercises a dominant influence. An approximate experimental relationship[24] is as follows:

$$\text{Microbend loss} \propto \left(\frac{\text{core radius}}{\text{fiber radius}}\right)^2 \cdot \left(\frac{1}{\text{NA}}\right)^4$$

Fiber Packaging

In order to reduce the influence of external forces, the fiber packaging design has been studied.[24] Based on the assumption that the most dominant type of external forces causing microbends are those caused by the fiber straddling a small dust particle or the roughness of the structures surrounding the packaged fiber which reappear as small bumps, a theory was formulated. This allows a comparative study of packaging arrangements of a two-layer structure of soft and hard material and the effectiveness of these arrangements against hydrostatic forces applied to the structure in the presence of surface roughness to be carried out.

ELASTIC DEFORMATIONS

When a fiber is pressed against an elastic plane surface with a certain degree of roughness (Fig. 3-13), the contact forces between the fiber and the surface are not uniform. The fiber will bend under the influence of the applied force $f(z)$ per unit length.

Using the theory of the thin elastic beam, the lateral displacement $x(z)$ of the fiber axis is related to $f(z)$ by

$$\frac{d^4x}{dz^4} = \frac{f(z)}{H}$$

where $H = EI$ is the flexural rigidity or stiffness, E is Young's modulus, and I is the moment of inertia.

On the basis of such a model, an analysis has been carried out which allows the comparison of the behavior of four different fiber coatings relative to their ability to minimize the microbend effects. The first coating is a soft plastic, the second is a hard plastic, and the third and fourth are hybrid structures. If it is assumed that the moduli for a typical soft material is 1 kg/mm² and 100 kg/mm² for a typical hard material, the excess loss can be computed for each structure for a mean lateral pressure which is taken as 0.1 g/mm. The rigidity (D) and stiffness (H) parameters used in each case are listed in the last two columns in Table 3-1. The decrease of the loss contribution

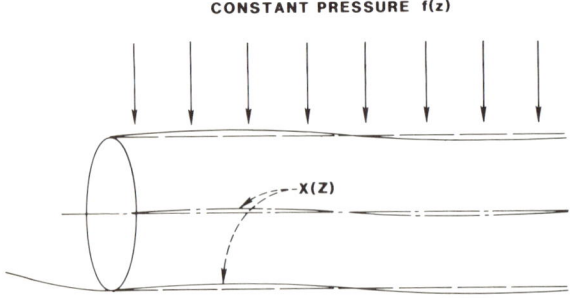

Fig. 3-13 Microbending of a fiber pressed with pressure $f(z)$ against a rough surface $x(z)$.

Table 3-1 Characteristics of Several Types of Protective Jackets for Optical Fibers

	Modulus, kg/mm²	Inside Radius, mm	Outside Radius, mm	Rigidity (D), kg/mm²	Stiffness (H), kg/mm²
Soft jacket	1	0.06	a_2	1	0.0713
Hard jacket	100	0.06	a_2	100	$0.0702 + 78.5(a_1)^4$
Inside jacket	100	0.06	$a_2 - 0.02$	1	$0.0702 + 78.5(a_2 - 0.02)^4$
Outside jacket	1	$a_2 - 0.02$	a_2		
Inside jacket	1	0.06	$a_2 - 0.04$	1	0.0713
Outside jacket	100	$a_2 - 0.04$	a_2	$1 + \dfrac{0.0064}{(a_2)^4}$	$78.5(a_2)^4 - 78.5(a_2 - 0.04)^4$

with increasing jacket radius in the case of the single material hard jacket results from the increase in overall stiffness.

In the case of a fiber with a hard outer coating over a soft inner coating, the hard outer coating provides the increased stiffness and can be made with a material which offers physical protection to the fiber from abrasion. The soft inner coating acts as a cushion to bending forces, thereby preventing microbends of small-radius curvature to be formed. A jacket diameter of 0.5 to 0.6 mm permits a virtual elimination of the distortion loss in the example considered here. A similar reduction by a single hard jacket requires at least twice this jacket diameter.

The ideal structure is a fiber surrounded by an infinitely soft material and then by a high-modulus material. This turns out to be a fiber within a hollow tube. In practice, the hollow tube creates significant problems: (1) the fiber may move within the tube; (2) fiber movement may damage the fiber surface; (3) the void may be subsequently occupied by undesirable liquids such as water; (4) the hard tube could buckle when bent sharply; or (5) the fiber assumes a helical position to accommodate jacket shrinkage at low temperatures.

A more attractive approach is to support the fiber with a low-modulus material initially, followed by a high-modulus material. This largely eliminates the aforementioned problems.

Physical Properties of Optical Fiber Waveguides

When optical fiber waveguides are used in a practical environment, their strength and durability are important. The mechanical requirements called for in a fiber cable design are more easily met if the fiber is strong and will not fail under stress during its operational life. To achieve a strong fiber, it is necessary to understand the strength and fatigue failure modes of glassy materials. It is also essential to understand the effects of relative humidity, temperature, and other environmental conditions.

STRENGTH OF GLASS

Glass is a strong, elastic, but nonmalleable or brittle material. Its ultimate strength is governed by the bonding forces within its material structure. The complexity of glass microstructures does not allow accurate determination of total strength by calculating the aggregate of bonding forces.

The practical strength is reached when the ultimate strength limit of a glass is exceeded at some portion of the structure. Glass strength is, therefore, dependent on the microheterogeneity. The existence of microheterogeneity in glass reduces its strength, since high stress concentration may occur locally. At the glass surface heterogeneity is prevalent as a result of the surface formation process and from physical damages.

THEORIES OF GLASS STRENGTH

Flawless Glass

The stress required to break a bond can be estimated for individual molecules.[25] The theoretical cohesive strength σ_t can be expressed as

$$\sigma_t^2 = \frac{2\gamma E}{8a}$$

where γ is the surface energy of material, E is Young's modulus, and a is the atomic spacing or bond length. For Si—O bonds, this gives $\sigma_t = 18 \times 10^3$ N/mm² or 18 GN/m² corresponding to a bond distance of 1.6 Å.

Stress-Induced Flaw

This theory assumes that flaw appears as a direct result of the application of stress.

One theory due to Cox[26] assumes that a certain fraction of the Si—O bonds in a glass are in a broken state at any given moment as a result of the statistical spread of vibrational energy. Since a broken bond is incapable of supporting a stress, an extra load will momentarily be placed on the bonds that are in the vicinity of the broken bond when a stress is applied. If a critical number of neighboring bonds are simultaneously broken, a fracture will propagate. Time-dependence, temperature-dependence, and moisture effects can all be predicted. This theory, however, is somewhat controversial.

Glass with Flaws

This theory starts by assuming that glass has flaws and concerns itself with the condition under which the flaw will propagate.[27]

It is assumed that the flaws have the geometric shape of narrow cracks with small radii of the curvature at their tips where applied stresses are concentrated. The stress at the tips can be calculated from elasticity theory. The calculation has been carried out for a straight crack of elliptical cross section with the applied stress perpendicular to the crack, as shown in Fig. 3-14.

The stress σ at the crack tip in the direction of applied stress S is

$$\sigma = S\left(1 + \frac{2L}{a}\right)$$

where L is the crack length (semimajor axis of the ellipse) and a is half the crack width (semiminor axis of the ellipse). With the radius of curvature $\rho = a^2/L$ at the crack tip and assuming $L \gg a$, σ is

$$\sigma = 2S\left(\frac{L}{\rho}\right)^{1/2}$$

The fracture stress σ is inversely proportional to the square root of the crack length. This dependence was found by Griffith in his well-known equation

$$\sigma = \left(\frac{2E\gamma}{\pi L}\right)^{1/2}$$

which gives the fracture stress for a crack of elliptical shape with semimajor axis length L and where E is Young's modulus and γ is the surface energy. The stress at the crack tip with load exerted along the minor axis of the ellipse is equal to the cohesive or bond strength.

For fibers with σ in the region of GN/m^2, the crack size responsible is of the order of 100 Å.

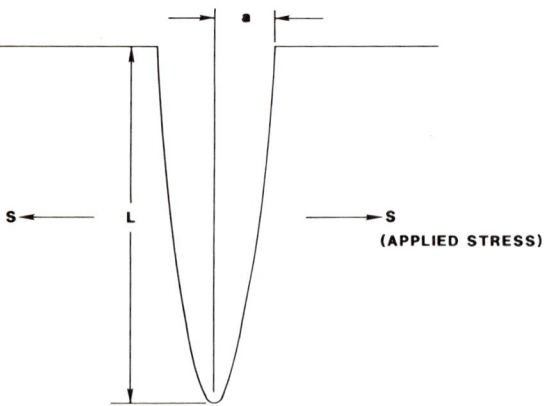

Fig. 3-14 Geometry of an elliptical flaw.

GLASS SURFACE STRUCTURES

Surface structure of oxide glasses depends mostly on the reaction of nonbridging oxygen at the surface during surface formation.[28] These bonds may react rapidly with atmospheric water to form silanol (SiOH) groups. The surface of a glass is, therefore, normally not free from metal hydroxyl groups. The structural arrangements depend on glass composition, thermal history, humidity, and surface treatment after melting and cooling.

The presence of physically absorbed molecular water is indicated by the presence of broad absorption bands at about 3450 cm^{-1} and another at about 1250 cm^{-1}. Other bands indicate that some hydroxyl groups are sufficiently close together to be hydrogen-bonded.

The density of isolated SiOH groups on a silica surface has been calculated as 1.4 groups per 100 Å and that of hydrogen bonded groups, as 3.2 groups per 100 Å. Figure 3-15 indicates the types of hydroxyl groups existing on silica surfaces. These are isolated silanol groups (Fig. 3-15a), two OH groups on one silica atom (Fig. 3-15b), and adjacent

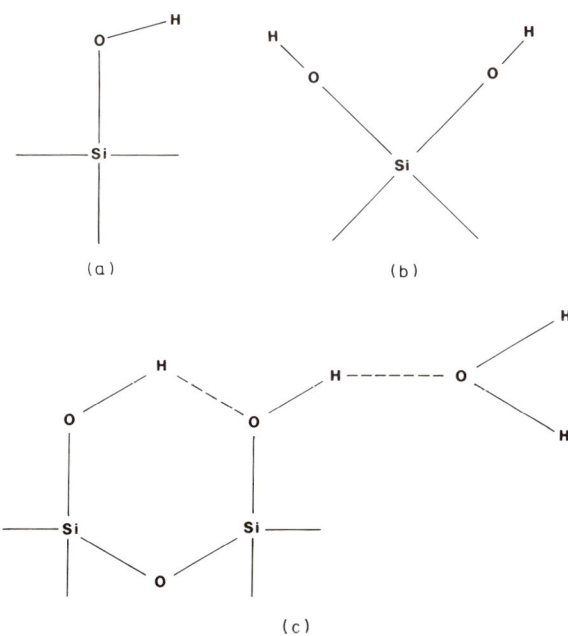

Fig. 3-15 Hydroxl groups found on silica surfaces: (a) isolated OH groups; (b) two OH groups on one silica atom; (c) hydrogen-bonded groups with an adsorbed water molecule.

OH groups bonded with hydrogen bond and molecularly absorbed water (Fig. 3-15c). The relative concentration of each group on silica surfaces depends on the melt atmosphere and thermal history of the glass and the temperature and humidity during the spectroscopic measurement.

Internal silanol groups near the surface have also been observed. These internal groups do not form at room temperature because of the low diffusion coefficient of water in bulk silica but can take place at elevated temperatures.

In the case of multicomponent glasses, other glass-formers will provide sites for hydroxyl groups. A monovalent cation R in silicate glass can undergo an exchange reaction with water as shown:

$$H_2O + SiO\text{—}R^+ \rightarrow SiOH + ROH$$

As hydrogen cations are smaller than the monovalent cations, the formation of O—H bonds at the glass surface can increase the reactivity due to the presence of residue stress and further hydration or crack formation in some cases.

FATIGUE STRENGTH

When a stress that is lower than the critical stress is applied in the presence of moisture, glass exhibits a time-delayed failure mode. Apparently, water, most likely in ionized form, attacks the glass surface and causes the formation of a weak bond which is broken by the applied stress. This initiates a progressive process, resulting first in a slow deepening of the flaw and then in the propagation of the flaw at progressively higher speed until the stress concentration reaches a critical value and fracture occurs. This is known as *fatigue*. The long-term strength of glass is lowered as a result of this phenomenon.

A theory due to Hillig and Charles[29] explains delayed fracture in glass in terms of stress-induced corrosion by water at flaw tips. The crack propagation was assumed to be an activated process in which the activation energy was stress-dependent. The basic equation for crack velocity can be developed from the absolute rate theory of chemical reactions. If the rate-limiting step for crack progagation is assumed to be an attack of the silicon-oxygen bonds by hydroxyl ions, the crack-propagation equation is

$$V = V_0[OH^-][A_g]\exp\frac{-\Delta E^{\ddagger} + \sigma\,\Delta V^{\ddagger}/3 - V_m\gamma/\rho}{RT}$$

where [OH⁻] is the hydroxyl ion activity at the crack surface and [A_g] represents the chemical activity of a flat glass surface in contact with the corrosive environment. The first term in the exponential $\Delta E\ddagger$ represents the activation energy for the chemical reaction. The stress at the glass-liquid interface is given by σ, and the activation volume for the chemical reaction is represented by $\Delta V\ddagger$. The final term in the exponential accounts for the changing chemical activity of the glass surface with surface curvature. The molar volume of the glass is given by V_m, γ is the interfacial surface tension at the glass medium interface, and ρ is the radius of curvature of the glass surface.

This is sometimes written in an alternative form

$$V = V_0 \exp(\beta\sigma)$$

where V is the velocity at which the glass surface corrodes under a tensile stress, V_0 is the rate of corrosion with no stress, and β is a constant parameter.

The experiments on the velocity of crack propagation at relatively low stresses verify this exponential form.[30] For a given system (environment, temperature, and glass composition), there is a unique relation between the crack velocity V and the crack-tip stress-intensity factor K_I. However, at higher stresses the velocity of crack propagation assumes different forms. Typically a trimodal curve is obtained. Two such curves are shown in Figs. 3-16 and 3-17.

At lower stresses (region I in Fig. 3-17) the velocity is exponentially dependent on stress. In this region the crack velocity can be expressed as a power function of the stress-intensity factor:

$$V = AK_I^n$$

where n and A are constant (typically $n = 15$ for glass). At higher stresses (region II) there appears to be a leveling off of the velocity

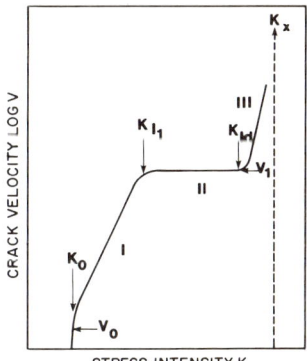

Fig. 3-16 Trimodal characteristics of crack propagation.

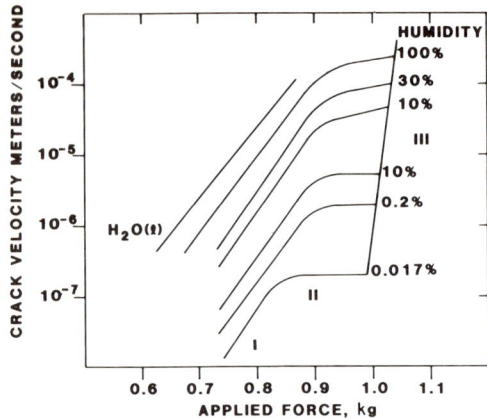

Fig. 3-17 Trimodal regions I, II, and III of a crack propagation at different humidities.

dependence on stress, until the velocity no longer depends on stress. In this transition region the velocity still depends on humidity. It is suggested that in this region the crack-propagation velocity is limited by the rate of transport of water vapor to the crack tip. At still higher stresses (region III) the velocity again increases exponentially with stress and is no longer dependent on the environment. For most glass systems, regions II and III occur at high velocities and the crack-propagation time in region I governs the time delay to failure.

For low-stress failures, the fracture surface of glass shows certain characteristics. The radial region closest to the initiating flaw is smooth and is known as the "mirror region." Beyond this region is a misty region where the surface roughens, and further away is a region of gross roughening known as "hackle." The small channels in the mist are the beginnings of branching of the crack, and the hackles are more extensive branchings of the crack.

Experimental evidences show that crack propagation ceases when stress is removed. Comparison of cyclic and static fatigue tests demonstrate that the time to failure depends on the magnitude and total duration of the applied stress, but not on its cyclic nature.

The data obtained from various experimenters show the following:

1. Crack growth at low stress is by and large exponentially dependent on the stress.
2. At a higher stress the stress dependence is almost absent. The rate of growth is suggested to be governed by the rate of transport of water to the crack tip.

3. At sufficiently high stress the propagation rate corresponds to the velocity of sound.
4. An acidic or basic environment is seen to influence the crack growth.
5. Crack propagation studies in vacuum indicate that for silica and some glasses, slow crack growth does not take place.
6. The factor n appears to be independent of the surface condition of the samples. This implies that the fatigue mechanism is probably a single process.

FIBER STRENGTH

It is reasonable to start by assuming that the glass properties in fiber form are not significantly different from those found in bulk or sheet forms, despite the greatly altered size and surface-to-volume ratio. Hence, the fiber strength is determined by the same mechanisms which govern glass strength. In particular:

1. Fracture originates from flaws (microcracks) (Fig. 3-18) formed as a result of (a) inhomogeneities, (b) phase separation arising from thermal conditions encountered during fiber fabrication, (c) interaction with the environment of the newly formed surface, (d) adsorbed ions, (e) crystallization, and (f) mechanical damage. The larger flaws are usually surface flaws caused by external contaminants or through mechanical abrasion. However, as *in situ* protection methods are perfected, the largest flaw may well be the result of the random surface structural differences. In order to maintain the pristine glass surface formed at the time of fiber drawing, a plastic is applied by a contactless method. The purpose of the protective jacket is not only to protect the external surface of the fiber, but also to

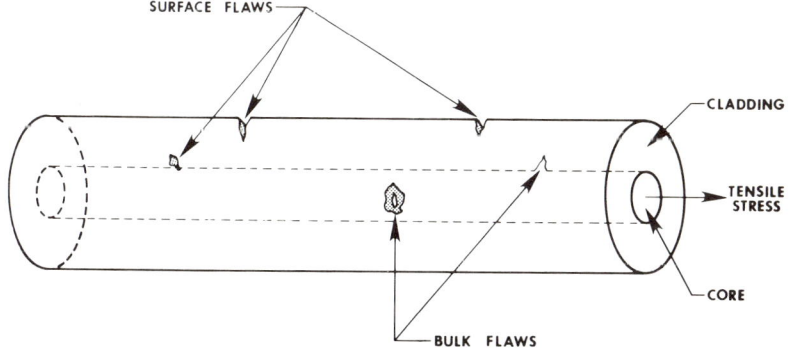

Fig. 3-18 Surface and bulk flaws along a fiber.

allow the fiber to be handled conveniently. This does not appear to modify significantly the pristine surface.
2. Under load, surface microcracks which are too small to cause catastrophic failure will enlarge or propagate. When a critical stress which equals the bonding force is reached at a crack tip, the crack will propagate at a high velocity, resulting in a fracture. The critical stress is most likely to be reached first at the tip of the deepest flaw since stress applied transversely to such a crack will produce a very high stress concentration at the tip of the crack through leverage action. Fatigue or stress corrosion is associated with the propagation of subcritical surface cracks under load in the presence of water or other reactants.
3. Internal stress within the fiber, caused by the differential thermal expansion coefficients of the different glasses making up the fiber, tends to put the core region in tension and the outer surface in compression, thus improving the strength.

FIBER DURABILITY

The existing theory of fatigue failure constructed from studies on bulk glass should be applicable to the fatigue phenomenon in fiber. Experimental investigations[31] of fiber durability have largely substantiated this belief. The pristine surfaces and the purity of the constituent material of the fiber have allowed more controlled studies to be carried out. It is expected that a more thorough understanding of the role of water in the stress corrosion process will eventually be resolved.

Much of the data gathered for fused silica optical fibers seem to fit Charles and Hillig's theory as applied to bulk fused silica glass.

First, the influence of water on static fatigue failure has been clearly demonstrated.[32] When the fiber is tested on a mandrel in air at room temperature, the n value obtained is about 24 (Fig. 3-19). When the same fiber is tested by complete immersion of the mandrel in water at room temperature, n changes (Fig. 3-20). At low stress levels n decreases to about 15, and beyond the threshold stress (3700 N/mm² or 525 ksi) n changes back to 24, the value found for tests in air. This indicates that:

1. Below 3700 N/mm² the increase in water concentration at the crack tip (from atmospheric water) involves a decrease of n from 24 to 15. This results in a decrease in time to failure as per the equation $\log t = -n \log A$, where A is a constant.
2. Above 3700 N/mm² there no longer seems to be any water-concentration dependence. However, other experiments have indicated that

it takes about an hour for water to penetrate through the plastic jackets. Since the time to failure at these stress levels is less than one hour, it is likely that a high concentration of water did not reach the glass prior to failure. For this reason, an n value similar to that of fibers tested in air was found. However, if the fiber is presoaked, $n = 15$ then prevails even at this high-stress region.

A similar experimental investigation[33] on uncoated fiber suggests that molecular H_2O does not attack SiO_2 directly and that OH^- ions play a crucial role in stress corrosion.

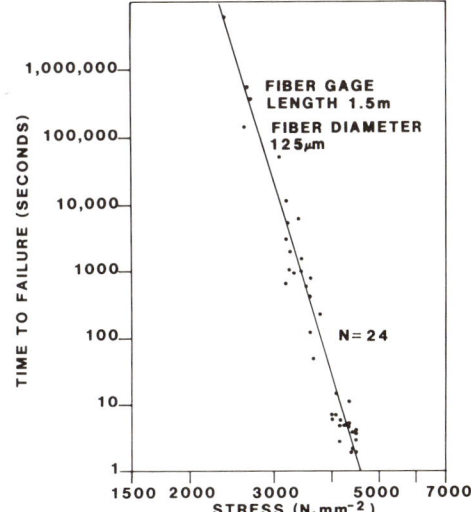

Fig. 3-19 Stress-induced time to failure in air of a fiber.

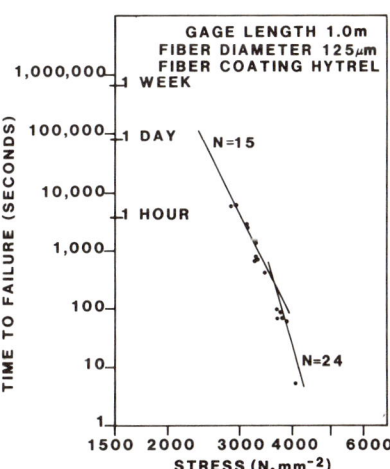

Fig. 3-20 Stress-induced time to failure in water of a plastic-coated fiber.

Activation Energy and Temperature Dependence

The activation energy for the chemical reaction can be deduced from the data gathered on fused silica optical fibers. One such experiment is to measure times to failure at different temperatures.[34] A typical set of results is given in Fig. 3-21. The activation energy calculation is derived from the slope of the straight line obtained by plotting $\ln(t)$ versus $(1/T)$. If λ is the slope of the line, then

$$\ln(t) = \frac{\lambda}{T} \quad \text{or} \quad t = t_0 \exp\frac{\lambda}{T}$$

Compared with the Arrhenius equation, λ can be identified as E/R, where E is the activation energy and R is the gas constant. As stress increases, the activation energy appears to decrease:

$E = 10.7 \text{ kcal/mol}$ at 3440 N/mm^2
$E = 11.1 \text{ kcal/mol}$ at 3200 N/mm^2
$E = 13.2 \text{ kcal/mol}$ at 2860 N/mm^2
$E = 18.0 \text{ kcal/mol}$ at 0 N/mm^2

Conclusions

The following evidences summarize the fiber aging characteristics:

- Water is determining factor affecting the aging characteristics of silica fibers.
- The number n is dependent on both OH^- concentration and temperature.

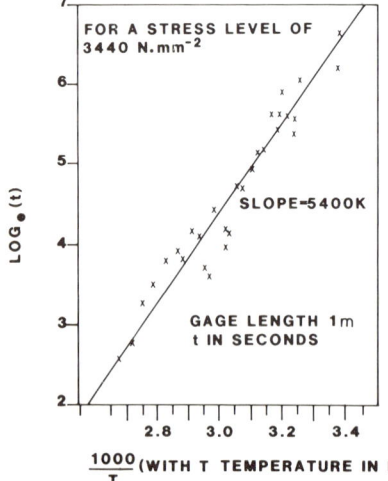

Fig. 3-21 Effect of temperature on time to failure.

- The activation energy of the water reaction with silica is a function of the stress applied to the fiber.

MINIMUM LIFETIME

The minimum lifetime of a fiber under load can be estimated by performing a proof test (see discussion of mechanical characteristics). During the proof test weak fibers are eliminated. The fibers passing the proof test will have a strength distribution not unlike that before the proof test, except that the distribution will be truncated at the proof-test stress level.

A design diagram has been constructed in Fig. 3-22 on the basis of minimum lifetime after proof-testing. Such a design diagram can be used to determine the required proof stress to guarantee a minimum lifetime in service. For a given proof-test stress, the allowable stress can be calculated for a given minimum lifetime. For example, to assure a minimum lifetime of 50 years under a constant applied stress of 1500 N/mm², the optical glass fiber would have to be proof-tested at a level of about 2.85 times the applied stress, or about 4275 N/mm².

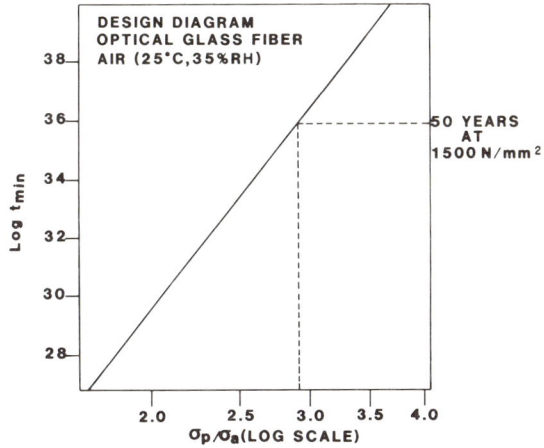

Fig. 3-22 Predicted life of a fiber under stress after proof test.

Fiber Evaluation Methods

The transmission characteristics and mechanical properties of optical fibers are measured by various techniques. The main transmission characteristics of interest are attenuation or loss, dispersion or delay and bandwidth. The principal mechanical characteristic of interest is the statistical distribution of fiber strength, which implicitly describes the durability of the fiber.

Measurements of Optical Characteristics of Fiber Waveguides

ATTENUATION MEASUREMENTS

An important and highly sensitive method is to measure the optical power transmitted through two lengths of a fiber with input launching and output detection conditions held invariant. A second technique employs an optical time domain reflectometer.

Insertion Loss

The optical power at a propagation distance X in the fiber waveguide can be expressed as

$$P(x) = P(x_1) \exp[-\int_{x_1}^{x} \alpha(x) dx]$$

where $\alpha(x)$ is the loss coefficient and $P(x_1)$ is the power in the fiber position x. If the loss coefficient is constant, the attenuation coefficient is given by

$$\alpha(x_1, x_2) = \frac{1}{x_2 - x_1} \int_{x_1}^{x_2} \alpha(x) dx$$

where x_1 and x_2 are the two positions along the fiber lengths used in the measurement. This loss equals the constant steady-state attenuation of the fiber, provided that each of the modes has the same attenuation. Individual modes, however, have different losses. High-order modes, which are more susceptible to coupling to unguided modes as a result of guide imperfections, suffer more loss than lower order modes. This differential loss tends to be statistically averaged out along the fiber by modal coupling. In fact, when complete averaging is achieved, a steady-state mode distribution is established which propagates with a characteristic attenuation constant.[11] However, the input launching condition, and the distribution and magnitudes of imperfections of a fiber, greatly influence the fiber length over which a steady-state mode distribution is established. Thus it is essential to specify the measurement conditions for accurate comparison of precise measurement results.

Optical Time Domain Reflectometry[35]

The loss measurement techniques discussed above provide the averaged insertion loss for a given fiber length but give no information concerning the length dependence of the loss. An optical time domain reflectometer can be used to measure attenuation and allows the length dependence of the fiber attenuation to be obtained, although at a reduced precision. This technique, based on the analysis of backscattered light in the fiber, requires neither cutting the fiber nor access to both ends of the fiber. It is, therefore, convenient for use with cabled fibers. In the backscattering experiments, a pulse of light is launched into the fiber in the forward direction by using either a directional coupler or a system of external lenses and beam splitter.

The waveform of the return light pulse is detected by a photodetector and processed with a boxcar integrator to improved signal-to-noise ratio (SNR). The return waveform consists of three distinct segments: (1) an initial pulse, which results from any lack of directionality in the input coupling mechanism; (2) a long tail caused by the distributed Rayleigh scattering that occurs as the input pulse propagates down the fiber; and (3) pulses caused by the discrete reflections along the fiber length as a result of fiber imperfections, in-line connectors, breaks, or the Fresnel reflection of the end of the fiber. The local slope of the Rayleigh-scattered energy-versus-fiber position can be used to extract the local attenuation coefficient. The time dependence of the detected backscattered power can be converted to a length dependence if the velocity of light in the fiber core is known. The detected backscatter power may be expressed as $p(x) - kp(0) \exp[-2\bar{\alpha}(x)x]$, where x is a position along the fiber length, k is a constant, $p(0)$ is the power launched into the fiber at the input, and $\bar{\alpha}(x)$ is the average total attenuation coefficient of the forward and backscattered signals.

DELAY DISTORTION

Delay distortion can be measured by one of several techniques, either in the time domain (impulse response measurements) or in the frequency domain (transfer function measurements). The impulse response is directly relevant for digital signals, while the transfer function is directly relevant for analog signals.

Single-Pass Impulse Response Measurements

A narrow pulse of light is injected into one end of a fiber, and the broadened output pulse is detected at the other end at a fast, sensitive,

58   Chapter Three

and linear detector. Computational facilities are required for deconvolving the received pulse obtained with the unknown fiber from the pulse that is received with a short fiber substituting for the unknown fiber.

Launch conditions are as important in dispersion measurements as in loss measurements. Since the launch affects the distribution of energy in the various modes, it affects the impulse response measurement. The impulse response should ideally be measured under various launch conditions and over different lengths[36] (different angles, spot sizes, etc.) to obtain a complete fiber characterization. In practice, however, this is too difficult and time-consuming to implement. Reasonable but conservative results can be obtained by exciting low-order modes only and measuring over a relatively short length (1 km).

Direct Measurement of the Transfer Function

As an alternative to using narrow pulses of high-intensity light, a sinusoidally modulated continuous-wave (CW) light signal can be used to directly measure the fiber transfer function. If the modulation frequency is swept and the transmission is monitored, the frequency at which the transmission level rolls off to 3 dB gives the bandwidth of the fiber. There are advantages to this technique; namely, the transfer function is obtained directly without Fourier-transforming time domain data. This technique has been used to measure the fiber transfer function at various wavelengths by using an incoherent light source, narrow-band optical filters, and an optical modulator to generate the sinusoidally varying light power.

In principle, both the amplitude and the phase characteristics of the transfer function are required to define the transfer function. But, in practice, only the amplitude variation is obtained. This means that phase distortions cannot be predicted from the bandwidth measurement results.

PROFILE MEASUREMENT

In a graded-index fiber the profile of the core refractive index controls the fiber bandwidth. An accurate determination of the profile can be used as a means to estimate the bandwidth performance.[37] Furthermore, the profile measurement is a diagnostic tool to monitor the accurary and repeatability of the profile achieved in the fabrication process. The profile can be measured both in the preform stage and in the fiber.

PREFORM PROFILE

The most reliable method[38] is to prepare a thin section of the preform and measure the index changes through an interference microscope. If the thickness of the section is uniform, this is an absolute method and can yield accurate results.

A second method is to illuminate the preform in a transverse direction with a coherent source.[39] The fringe pattern is a measurement of the profile.

FIBER PROFILE

A transverse illumination method[39] can also be used to give the profile. This method can be used during the fiber-drawing process and is a useful on-line measurement technique. The accuracy, however, is limited.

A precise method of measuring fiber profile of a short length of fiber is based on the variation of the local numerical aperture of the graded-index fiber. A launching spot scanned across the fiber with a numerical aperture higher than the maximum fiber numerical aperture ensures that radiative modes are excited along with the propagating and leaky modes.[40] At the output end only the radiative modes are detected. This method yields index distribution along the line of the scan.

Other methods include measurement of the near field[41] or far field[42] when the input is excited in a prescribed manner, and measurement of local reflection at the input end when the fiber is illuminated with a spot beam. These methods must be used with caution since they are susceptible to errors arising from leaky modes and surface contaminations, respectively.

Measurement of Mechanical Characteristics of Fiber Waveguides

FIBER STRENGTH MEASUREMENT

Fiber strength is measured in order to determine the statistical variation of fiber strength along the length of a fiber (due to the spatial distribution of flaws). Test procedures which enable the prediction of reliable long-length strength are desirable; the procedures should be simple and should not involve the destructive test of a large amount of fiber. If sufficient data can be gathered such that the statistics of the large but rare flaws are characterized, then prediction in statistical terms of the tensile strength of a particular gauge length of fiber can be defined with known confidence limits.

STRENGTH TEST PLAN

The mechanical strength test plan for high-strength fiber is a systematic approach to gather statistically relevant data. The procedures of performing a series of strength tests and reduction of data obtained from these tests are defined, so that fiber strength may be determined and appropriately specified. The test plan calls for dynamic fatigue tests, static fatigue tests, and proof tests. Data are gathered and processed (using Weibull or other suitable statistics) to provide a coherent and correlated set of results from which fiber strength specifications may be drawn.

Dynamic Fatigue Test

The dynamic load test measures the mechanical tensile strength of a fiber of a chosen gauge length at a defined constant rate of loading. The average strength and strength distribution of a fiber are recorded. For a high-strength fiber, the fiber strength spread is expected to be small and essentially unimodal. These tests could be carried out with a specimen gauge length of around 1 m. In order to obtain statistically relevant data, at least 20 samples, and preferably 40 samples, should be tested. Different rates of loading spanning several orders of magnitudes, e.g., 0.2 to 200 N/mm^2, would provide the required data. This test is meaningful only for fibers with no rare large flaws and provides information on the basic fiber strength.

The test equipment is illustrated in Fig. 3-23. The fracture load is applied through a constant-speed drive motor at a constant rate until

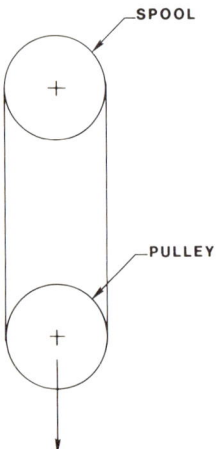

Fig. 3-23 Dynamic fatigue test apparatus.

fracture occurs. The load is recorded, and the diameter of the broken fiber end is measured and recorded.

Static Fatigue Testing

To investigate time to failure versus stress for fibers without resorting to statistical testing, a static fatigue test method involving the application of stress through bending is convenient. As the test is carried out over a small section of the fiber, the probability of encountering a rare large flaw is extremely low. The fatigue characteristics measured are those corresponding to a fiber with no rare flaws.

Bending Stress

The test equipment consists of precision mandrels of different diameters or simply by bending a loop of fiber to different radii. The stress level can be varied by the proper choice of the mandrel size.

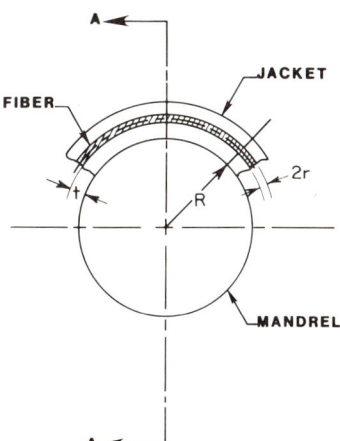

Fig. 3-24 Geometry of a fiber over a mandrel.

Take R the radius of the mandrel, r the radius of the fiber, and t the thickness of the plastic cladding (Fig. 3-24), and calculate the elongation at the outermost portion of the core surface as follows. If L_0 is the neutral axis length and L is the maximum length of the fiber under bending stress, the elongation is $(L - L_0)$. Hence the strain is

$$\epsilon = \frac{L - L_0}{L} = \frac{(R + t + 2r)\,\theta - (R + t + r)\,\theta}{(R + t + r)\,\theta} = \frac{r}{R + t + r}$$

and the stress

$$\sigma = \epsilon E = \frac{Er}{R+t+r}$$

with $E =$ Young's modulus (typically 7.2×10^{10} N/mm² for fused silica).

Convenient mandrel sizes for testing and the corresponding flexual stresses are shown in Table 3-2 for a fiber diameter of 125 μm (5 mil).

Table 3-2 Equivalent Stresses for Different Mandrel Diameters

Mandrel Diameter, mm	Equivalent Stress, N/mm² for 125-μm fiber diameter	Equivalent Percent Elongation
1.55	4152	6.0
2.08	3311	4.8
2.38	2972	4.3
2.73	2655	3.8
3.13	2365	3.4

Note: Equivalent stresses induced by bending.

Proof Test

The minimum strength of a long length of fiber is governed by the largest flaws of the weakest point along the entire length of the fiber. In the proof test the entire fiber is subjected to a proof stress with a view to ensure that the fiber passing the test will have a minimum guaranteed strength equal to the proof stress. In order to minimize the effect of stress corrosion, the load is applied to a small length of the fiber for a time which should be kept to a minimum.[31] In the region where the load is applied, the fiber is held on the two drive units to prevent any slip. The tension is set by using adjustable torque drive or by using a dead weight. The stress is completely relieved after the test through the use of low-tension take-up as the fiber completes its testing period. The setup should be fitted as part of the fiber-drawing apparatus, since the freshly made fiber can be protected readily from moisture until the proof testing has been completed. This would ensure that the flaw distribution after the proof test remains unchanged except the truncation of the flaws which will fail at the proof stress.

Fiber-Fabrication Processes

A number of different fiber-fabrication techniques are to be described, in general terms, under the following descriptive headings:

- External chemical vapor deposition of soot (external CVD)
- Internal chemical vapor deposition of glass (internal CVD)
- External chemical vapor deposition of glass (plasma CVD)
- Multielement glass
- Phasil system

EXTERNAL CHEMICAL VAPOR DEPOSITION

External CVD is a batch process which has continuous operation possibilities. This process produces core and cladding materials in ultrapure form. However, care must be taken to exclude externally induced contaminations. This method is capable of producing both multimode step and graded-index fibers of high quality but is less convenient for producing a single-mode fiber.

Process 1. The glass of the desired composition is deposited via a flame hydrolysis process in the form of a powder or soot, layer by layer, uniformly over the length of a mandrel. When deposition is completed, the mandrel is removed and the material is sintered.[43] The consolidated tube is then fused and collapsed simultaneously into a preform rod. The principles are illustrated in Fig. 3-25.

Process 2. The glass of the desired composition with or without radial variation is deposited via a flame hydrolysis process in the form of a powder or soot over the end of a starting rod.[44] The deposition is continued while the rod is slowly moved away from the flame. A soot boule will be formed in that direction. The soot is dried, sintered, and fused at a convenient position some distance away from the deposition end. This results in the continuous formation of a preform rod. The deposition schematic is as shown in Fig. 3-26.

With the use of readily available glass-forming materials such as

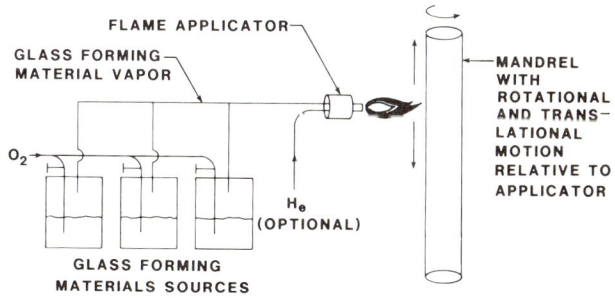

Fig. 3-25 Schematic illustration of an apparatus for external chemical deposition.

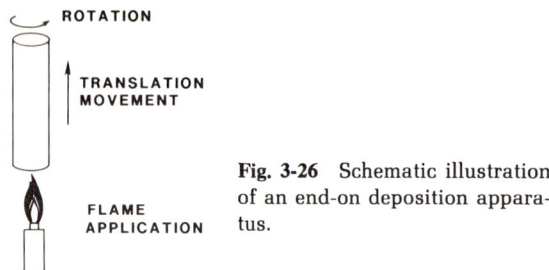

Fig. 3-26 Schematic illustration of an end-on deposition apparatus.

germanium, boron, phosphorus, fluorine, and silicon compounds, the processes can achieve the following performances:

1. Typical deposition rate: 1 to 2 g/min.
2. Loss: no absorption loss other than that due to small amount of OH^-, plus scattering loss; OH^- extraction is possible.
3. Typical NA: 0.2.
4. Dimensional restriction: none.
5. Dispersion: 600 MHz·km readily achieved; 1 Ghz·km possible.
6. Strength: adequate.
7. Dopant material: germanium, boron, phosphorus, fluorine.
8. Large preform size.

INTERNAL CHEMICAL VAPOR DEPOSITION

Internal CVD is a batch process. It forms the core and the cladding materials in ultrapure state within a substrate tube. It is capable of producing both single-mode and multimode step and graded-index fibers of high quality.

Process 1. The glass of the desired composition is deposited layer by layer from a vapor-phase reaction in the form of soot and fused immediately into glass uniformly over the length of the internal surface of the substrate tube[45] (Fig. 3-27). The heat required to induce the chemical reaction and to fuse the deposited material into a glass is supplied by an external heating source.

1. Typical deposition rate: 0.5 g/min.
2. Loss: scatter loss limited.
3. Typical NA: 0.22.
4. Dimensional restriction: outer diameter (OD)-to-core diameter ratio unlikely to be less than 1.5:1.

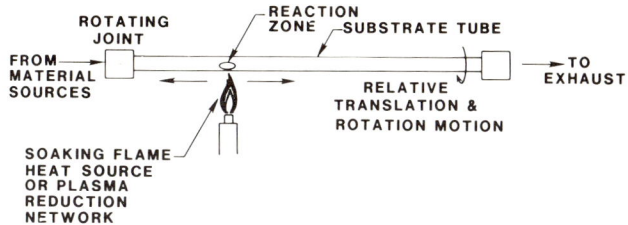

Fig. 3-27 Schematic illustration of an apparatus for internal chemical vapor deposition.

5. Dispersion: 1 GHz·km possible; 600 MHz readily achieved.
6. Strength: adequate.
7. Dopant material: germanium, boron, phosphorus, fluorine.
8. Preform size: 10 to 20 km of 125-μm-OD fiber.

Process 2. The heat of reactions is generated by electric current heating of the ionized gases in the form of an isothermal plasma (Fig. 3-28).[46] The deposited material is in glassy form. The gases are flowing at a rate similar to the flow rates of gases in the indirect heating case described in process 1. The plasma temperature is extremely high, but the tube temperature is below the glass-softening temperature. The deposition rate can be high, while the tube is more stable during deposition, and hence scaling up to larger preform should be more readily achieved. However, at a high deposition rate the fiber quality achievable is impaired.

1. Typical deposition rate: 1.0 g/min.
2. Loss: scattering loss limited.
3. Typical NA: 0.2.
4. Dimensional restriction: as in process 1.
5. Dispersion: 600 MHz·km possible.

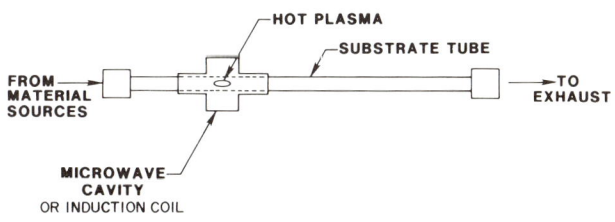

Fig. 3-28 Schematic illustration of an apparatus for internal chemical vapor deposition in which the heating is provided by a plasma region.

6. Strength: adequate.
7. Dopant material: germanium, boron, phosphorus, fluorine.
8. Preform size: 10 to 20 km of 125-μm-OD fiber.

Process 3. The heat of reactions is generated by electric heating of the ionized gases at reduced pressure in the form of a nonisothermal heterogeneous plasma,[47] and glass is deposited directly. The reaction efficiency is 100 percent, but the quantity of material flow is limited. The plasma temperature is well-controlled. Deposition is made on the inner surface of a substrate tube heated to below softening point. The deposition rate is low but can be extremely precisely controlled. The tube is stable during deposition. Scaling up is possible lengthwise but not easily in diameter. The arrangement is similar to the one shown in Fig. 3-28.

1. Typical deposition rate: 0.2 g/min.
2. Loss: scatter loss limited.
3. Maximum NA: 0.2.
4. Dimensional restriction: as in process 1.
5. Dispersion: 1 GHz·km possible.
6. Strength: adequate.
7. Dopant material: germanium, boron, phosphorus, fluorine.
8. Preform size: 10 km of 125-μm-OD fiber.

EXTERNAL CHEMICAL VAPOR DEPOSITION OF GLASS

Plasma CVD is the most highly developed process for making synthetic silica on an industrial scale.[48] The addition of a dopant, however, gives rise to the need to control the differential vaporization characteristics. This process has been successfully employed to make a fluorine and boron doped silica. Very large boules can be fabricated. The arrangement is similar to that shown in Fig. 3-26, except that the flame torch is replaced by a plasma torch.

1. Typical deposition rate: 1.0 g/min.
2. Loss: scatter loss limited.
3. Maximum NA: 0.2.
4. Dimensional restriction: none.
5. Dispersion: 300 GHz·km marginally achievable.
6. Dopant material: fluorine, boron, germanium, phosphorus.
7. Maximum preform size: unlimited.

MULTIELEMENT GLASSES

Multielement glasses are made from ultrapure basic oxides and carbonates.[49] The purification of raw chemicals is carried out by wet chemical processes. To avoid recontamination, the chemicals are mixed in a clean environment and fired in pure silica crucibles. The resultant glasses are drawn from crucibles in rod form ready for remelting in double crucibles for fiber formation (Fig. 3-29).

The purity achievable is not as high as those produced by the CVD process, since the wet chemical methods are less effective in separating out the transition elements. (In the CVD processes the transition-metal halides have very low vapor pressure and hence are not carried into the reaction zone.) The residue transition-metal ions give rise to nonnegligible absorption losses, unless steps are taken to shift the absorption bands of these transition elements out of the spectral region of interest. This can be done by controlling the oxidation or reduction conditions imposed during glass melting.[50]

The use of multielement glass is a natural candidate for large-volume continuous fiber fabrication and has cost advantages for large-scale production. Furthermore, the range of glass compositions is potentially larger.

1. Production rate: very high.
2. Loss: total loss 3 dB/km at 0.85 μm achievable for some glass systems. Poor long-wavelength performance as a result of OH^- absorption.
3. Numerical aperture range: 0.2 to 0.6.

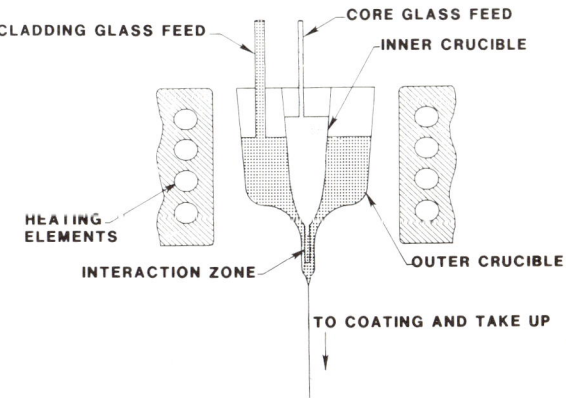

Fig. 3-29 Schematic illustration of a double-crucible arrangement.

4. Dimensional restriction: none.
5. Dispersion: 300 MHz·km.
6. Strength: depends on glass systems but in general is lower and less durable than silica fibers.
7. Composition: Na-B-Si; K-B-Si; Ge-B-Si; Na-Ca-Si; K-B-Si; Th-B-Si; Pb-Si.

PHASIL SYSTEM

The Phasil system is a fabrication process based on the Vicor glass process.[51] It is a batch process. The steps involved are lengthy but can be made efficient by processing a large number of preforms in parallel. The Vicor rods are first leached so that only a honeycomb silica structure is left. Then the desired pure dopant molecules are stuffed into the rod by a solution treatment. The clad layer is formed by a second partial leaching process. The structure is dried and sintered in order to form the fiber preform. By selective stuffing and leaching, profiled core can also be achieved. Figure 3-30 is a flowchart of the Phasil process.

1. Production rate: high.
2. Loss: 5 to 10 db/km, caused by absorption and scattering.
3. Typical NA: 0.2.
4. Dimensional restriction: none.
5. Dispersion: 300 MHz·km possible.
6. Strength: showing existence of internal flaws.

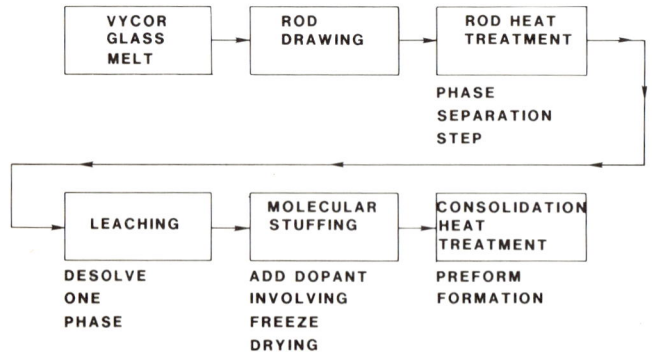

Fig. 3-30 Block schematic of the Phasil process.

COMPARISON OF FABRICATION TECHNIQUES

The five fiber-fabrication techniques can be ranked as follows:

Internal CVD—best performance but expensive.
External CVD—good performance, less expensive.
Plasma CVD—moderate performance, moderate cost, at moderate volume.
Phasil—moderate performance, low cost, at large volume.
Multielement glass—moderate performance, low cost, at very large volume.

Of these, the multielement glass method produces glasses which are to be remelted in a double-crucible structure to produce fibers possibly in continuous lengths. The others yield fiber preforms in a rod form. These rods are to be drawn into fibers in a batch mode. However, if the preform is sufficiently large, the loss of efficiency and material wastage due to setting up requirements may be minimal. The fiber loss specifications expected from these processes are compared in Fig. 3-31. It is to be noted that high-silica glass has very promising low-loss characteristics at the wavelength region 1 to 1.6 μm, while multielement glasses tend to have the loss minimum at the 0.8- to 0.9-μm region.

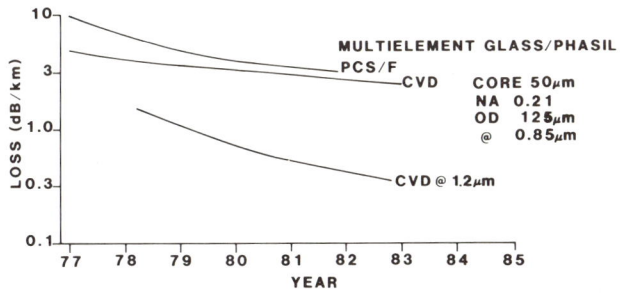

Fig. 3-31 Attenuation specification improvement with time.

Fiber-Drawing Processes

The double crucible is a fiber-drawing technique where fiber is drawn from the melts of the core and cladding glasses within a concentric crucible with central orifices. Another technique consists of heating the fiber preform tip and elongating the molten glass at the tip to form the fiber. Various heating arrangements can be used.

DOUBLE CRUCIBLE

The double crucible, as shown in Fig. 3-29, consists of a set of two concentrically placed platinum crucibles with nozzles at the centers of the crucible bases. The inner crucible contains the core glass which flows through the outer crucible containing the cladding glass. The nozzles are arranged with such spacing and of such sizes as to allow the desired diffusion to take place between the core and cladding glasses and to yield the desired core and cladding dimensions as the core glass flows through the cladding glass. The crucibles must not be a source of contamination and hence must be made from ultrapure platinum and be preleached. The crucibles are heated by induction or direct current or in a muffled furnace. The operating temperature is kept to a minimum. A maximum operating temperature is about 800°C; thus limiting the types of glass systems usable.

For continuous operation, the crucibles are kept at constant levels by feeding glass canes into the melt from the top of the crucibles at a suitably controlled rate. The maximum fiber-drawing rate is at least several hundred meters per minute.

FIBER DRAWING FROM PREFORMS[52]

Fiber drawing from fiber preforms involves heating the preform tip. Several heating sources are suitable, but each has its limitations.

Oxyhydrogen Flame. Flame stability within the combustion zone depends on torch design and intertorch interference. A stable flame is required, since fluctuations result in uneven heating and cause the fiber produced to have dimensional variations. With a well-designed flame, achievable tolerance is about several percent. The volume of heat available from the flame can be made sufficiently large for heating large preforms. The flame shape can be designed to avoid production of intense local heating spots which can cause preform vaporization. A drawing speed of 60 m/min is readily achieved. The strength of the fiber produced is excellent. It is to be noted that optical and physical properties of the fiber depend on the temperature and viscosity of the glass during the elongation process.

Furnaces. A graphite furnace can be energized by direct current or by alternating current through induction. The operating temperature required for silica fiber drawing is sufficiently high to cause silica and carbon to interact. In order to avoid contamination, the furnace design is critical. Suitable shielding and purging gas are usually necessary. Few furnaces are capable of producing sufficient heat for handling large preforms. Drawing speed of 60 m/min for 11-mm-diameter preform is readily achieved. With feedback control, 1 percent tolerance is easily accomplished. High strength is produced if care is taken to eliminate sources of contamination through purging.

An inductively heated zirconium furnace is another example of a high-temperature furnace for silica fiber drawing. The heat capacity and contamination control are good.

The carbon dioxide laser is another heating source for silica fiber drawing. Unfortunately, the absorption coefficient of silica of the 10.6-μm radiated power from the laser is extremely large, and the heat conductivity of silica is relatively low. As a result, the fiber preform surface reaches extremely high temperature and evaporates rapidly, thus limiting the size of preform which can be drawn into fiber by using laser heating.

FIBER COATING[52]

The as-drawn fiber has extremely high strength. However, the pristine fiber surface must be immediately protected from abrasion and interaction with the environment. One technique is to coat the fiber with a plastic coating by a solution coating method.

Ultraviolet (UV)-cured epoxy and room-temperature vulcanized (RTV) silicone are solventless solutions and can be rapidly cured. They are suitable and convenient as fiber-coating materials. Application techniques included the use of slightly tapered solid nozzle and flexible nozzle made from plastic. The aim is to coat the fiber without damaging the fiber surface through inadvertent contact of the fiber surface with the nozzle. Thus the nozzle must be designed to provide a certain amount of self-centering forces in order to achieve contactless coating and to permit the formation of a uniform concentric structure.

Solution coating speed is limited by rheological properties of the liquid, rate of curing, and the cooling rate of the fiber as it leaves the hot zone. The upper speed limit can be expected to exceed 300 m/min.

The critical parameters for fiber drawing are summarized in the following list:

1. Drawing temperature: (a) silica/doped silica; (b) multielement glass.
2. Heating method: (a) graphite furnace or zirconia furnace (both furnaces can be either dc or inductively powered); (b) CO_2 laser; (c) O_2—H_2 flame; (d) crucible heating.
3. Coating material: (a) epoxy; (b) silicone RTV; (c) UV-cured epoxy.
4. Preform size.
5. Drawing rate.

References

1. R. E. Collin, *Field Theory of Guided Waves*, McGraw-Hill, New York, 1960.
2. D. Hondros and P. Debye, "Electromagnetic Waves along Long Cylinders of Dielectric," *Annaleu der Physik* **63**(7):645–673 (1910).

3. M. Born and E. Wolf, *Principles of Optics*, 3d ed., Pergamon Press, New York, 1965.
4. A. Ankiewcz and C. Pask, "Geometric Optics Approach to Light Acceptance and Propagation in Graded Index Fibers," *Opt. Quantum Electron.* **9**:87–109 (1977).
5. D. Marcuse, *Light Transmission Optics*, Van Nostrand Reinhold, Princeton, N.J., 1972.
6. A. W. Snyder, "Asymptotic Expressions for Eigenfunctions and Eigenvalues of a Dielectric or Optical Waveguide," *Transact IEEE* **MTT-17**:1130–1138 (1969).
7. D. Gloge, "Weakly Guiding Fibers," *Appl. Opt.* **10**:2252–2258 (1971).
8. D. Marcuse, *Theory of Dielectric Optical Waveguides*, Academic Press, New York, 1974.
9. J. Sakai and T. Kimura, "Bending Loss of Propagation. Modes in Arbitrary-Index Profile Optical Fibers," *Appl. Opt.* **17**(10):1499–1506 (1978).
10. K. Okamoto, T. Edahiro, A. Kawana, and T. Miya, "Dispersion Minimization in Single Mode Fibers over a Wide Spectral Range," *Electron. Lett.* **15**(22):729–731 (1979).
11. S. D. Personick, "Time Dispersion in Dielectric Waveguide," *Bell Syst. Tech. J.* **50**:843–859 (1971).
12. S. Kawakami and J. Nishizawa, "An Optical Waveguide with the Optimum Distribution of the Refractive Index with Reference to Waveform Distortion," *Transact. IEEE* **MTT-16**:814–818 (1968).
13. R. Olshansky and D. B. Keck, "Pulse Broadening in Graded-Index Optical Fibers," *Appl. Opt.* **15**:483–491 (1976).
14. I. P. Kaminow and H. M. Presby, "Profile Synthesis in Multicomponent Glass Optical Fibers," *Appl. Opt.* **16**:108–112 (1977).
15. R. D. Maurer, "Light Scattering by Glasses," *J. Chem. Phys.* **25**:1206–1209 (1956).
16. D. Marcuse, "Derivation of Coupled Power Equations," *Bell Syst. Tech. J.* **51**:229–237 (1972).
17. D. Marcuse, "Coupled Mode Theory of Round Optical Fibers," *Bell Syst. Tech. J.* **52**:817–842 (1973).
18. D. Marcuse, "Pulse Propagation in Multimode Dielectric Waveguides," *Bell Syst. Tech. J.* **51**:1199–1232 (1972).
19. C. Yeh, "Modes in Weakly Guiding Elliptical Optical Fibers," *Opt. Quantum Electron.* **8**:43–47 (1976).
20. R. B. Dyott, J. R. Cozens, and D. G. Morris, "Preservation of Polarisation in Optical Fibre Waveguides with Elliptical Cores," *Electron. Lett.* **15**(13):380–382 (1979).
21. R. H. Stolen, V. Ramaswamy, P. Kaiser, and W. Pleibel, "Linear Polarisation in Birefringent Single-Mode Fibers," *Appl. Phys. Lett.* **33**(8):699–701 (1978).
22. E. A. J. Marcatile, "Bends in Optical Dielectric Guides," *Bell Syst. Tech. J.* **48**:2103–2132 (1969).
23. W. B. Gardner, "Microbending Loss in Optical Fibers," *Bell Syst. Tech. J.* **54**(2):457–465 (1975).
24. D. Gloge, "Optical Fiber Packaging and Its Influence on Fiber Straightness and Loss, *Bell Syst. Tech. J.* **54**(2):245–261 (1975).
25. R. H. Doremus, *Glass Science*, Wiley, New York, 1973, Chapter 15, pp. 281–283.
26. S. M. Cox, "Glass Strength and Ion Mobility," *Phys. Chem. Glasses* **10**(6):286–289 (1969).
27. A. A. Griffith, "The Phenomena of Rupture and Flow in Solids," *Phil. Transact. Roy. Soc.* **A221**:163–198 (October 21, 1920).
28. B. E. Warren, "Surface Structure of Glass," *J. Am. Ceramic Soc.* **21**:259–265 (1938).
29. W. B. Hillig and R. J. Charles, *High Strength Materials*, V. F. Zackey, ed., Wiley, New York, 1965, pp. 682–705.
30. S. M. Wiedenhorn, "Environmental Stress Corrosion Cracking of Glass," *Corrosion Fatigue NACE* **2**:731–742 (1972).

31. J. E. Ritter, Jr., J. M. Sullivan, Jr., and K. Jahus, "Application of Fracture-Mechanics Theory to Fatigue Failure of Optical Glass Fibers," *J. Appl. Phys.* **49**:4779–4782 (1978).
32. C. Kao, "Optical Fibre and Cables," in *Optical Fibre Communications*, M. J. Howes and D. V. Morgan, eds., Wiley, New York, Chapter 5, pp. 189–249.
33. A. Koa, "The Fatigue Strength of Uncoated Silica Fibres in Air and Solution of HCl and NaOH" (private communication), February 15, 1980, pp. 1–10.
34. C. Kao, M. Maklad, and V. Schurr, "Fatigue Strength of Strong Fibers at High Temperatures," Williamsburg Meeting on Optical Transmission, Williamsburg, Va. (February 21–24, 1977), paper TUA-6, pp. 1–4.
35. M. Barnoski and S. Personick, "Measurements in Fiber Optics," *Proc. IEEE* **66**(4):429–441 (1978).
36. L. G. Cohen, P. Kaiser, J. McChesney, P. B. O'Connor, and H. M. Presby, "Transmission Properties of a Low Loss Parabolic Index Fiber," *Appl. Phys. Lett.* **26**:472–474 (1975).
37. H. M. Presby, D. Marcuse, and L. C. Cohen, "Calculation of Bandwidth from Index Profiles of Optical Fiber, Pt. II, Experiments," *Appl. Opt.* **18**(19):3249–3255 (1979).
38. C. A. Barrus and R. D. Standley, "Viewing Refractive Index and Small Scale Inhomogeneities in Glass Optical Fibers: Some Techniques," *Appl. Opt.* **13**(10):2365–2369 (1974).
39. H. M. Presby and D. Marcuse, "The Index-Profile Characterisation of Fiber Preforms and Drawn Fiber," *Proc. IEEE* **8**(10):1198–1203 (1980).
40. W. J. Stewart, "A New Technique for Measuring the Refractive Index Profiles of Graded Optical Fibers," *Proc. IOOC (Tokyo)*: 395–398 (1977).
41. F. M. E. Sladen, D. N. Payne, and M. J. Adams, "Determination of Optical Fiber Refractive Index Profiles by a Near Field Scanning Technique," *Appl. Phys. Lett.* **28**(5):255–258 (1976).
42. M. J. Adams, D. N. Payne, and F. M. E. Sladen, "Length Dependent Effects Due to Leaky Modes on Multimode Graded Index Optical Fibers," *Opt. Commun.* **17**(2):204–209 (1976).
43. P. C. Schultz, "Method of Forming A Light Focusing Fiber Waveguide," U.S. Patent 3,826,560 (issued July 30, 1974).
44. T. Izawa and N. Inagaki, "Materials and Processes for Fiber Preform Fabrication-Vapor-Phase Axial Deposition," *Proc. IEEE* **68**(10):1184–1187 (1980).
45. J. B. MacChesney, R. E. Jaeger, D. A. Pinmow, F. W. Ostermayer, T. C. Rich, and L. G. Van Uitert, "Low Loss Silica Core-Borosilicate Clad Fiber Optical Waveguide," *Appl. Phys. Lett.* **23**:340–341 (1973).
46. K. Fujiwara, N. Yoshioka, M. Hoshikawa, T. Miyeshita, and H. Takata, "Optical Fiber Fabrication by Isothermal Plasma Activated Deposition," *Proc. 3d ECOC* (European Conference on Optical Communications), *Munich:* 15–17 (1977).
47. D. Kuppers and J. Koenings, "Preform Fabrication by Deposition of Thousands of Layers with the Aid of Plasma Activated CVD," *Proc. 2d ECOC, Paris:* 49–54 (1976).
48. A. Müklich, K. Ran, and N. Treber, "Preparation of Flourine Doped Fused Silica Preform by Plasma Chemical Techniques," *Proc. 3d ECOC, Munich:* paper No. 59, 10–11 (1977).
49. K. J. Beales, C. R. Day, A. G. Dunn, and S. Partington, "Multicomponent Glass Fibers for Optical Communications," *Proc. IEEE* **68**(10):1191–1194 (1980).
50. K. J. Beales, C. R. Day, W. J. Duncan, and G. R. Newns, "Low Loss Compound-Glass Optical Fiber," *Electron. Lett.* **13**:755–756 (1977).
51. P. Macedo, J. H. Simmons, T. Olsen, R. K. Mohr, M. Samanta, P. K. Gupta, and T. A. Litovitz, "Molecular Stuffing of Phasil Glasses for Graded Index Optical Fibers," *Proc. 2d ECOC, Paris:* 37–39 (1976).
52. L. L. Blyler and F. V. DiMarcello, "Fiber Drawing Coating and Jacketing," *Proc. IEEE* **68**(10):1194–1198 (1980).

Chapter

Fiber Cables

A fiber with a protective coating is like a copper wire with its primary insulation. It can be conveniently handled, is basically strong, and can survive in a relatively benign environment. For the fiber to be used in different environments and to fulfill different roles, it must be suitably incorporated into cable structures. Many fiber cable types are required for different applications. They range from a single-fiber, lightweight, general-purpose cable intended for use as an interconnect wire, to a multifiber cable for carrying signal traffic between telephone switching offices via underground ducts, to an armored marine cable designed to provide a communications channel between the ship and the vehicle it is towing with the cable. Obviously, a great variety of operational conditions exist. The cables must provide an environment for the fiber to survive and perform when the cable is exposed to the full operational conditions.

The principal design constraints are governed by the physical and optical performance requirements. In general, the design alternatives are narrowed if the cable cross-sectional dimension, weight, and strength are restricted. A successful fiber cable must maintain its optical performance over the designed mechanical and environmental limits. This implies that the cable structure must not exert excessive lateral forces to increase microbending losses of the fiber and must not permit the fiber to be subjected to transient tensile stresses over a designed maximum value or an average steady tensile stress over the fatigue limit, which could shorten the fiber service life.

The choices of the structural geometry, the strength-rendering material, the fiber arrangement, and the supporting material offer relatively wide design alternatives. The structural geometry should offer ease of storage and laying, longitudinal flexibility, and circumferential rigidity. This favors a circular cross section. The strength-rendering material provides the additional strength to the cable. Its choice greatly influences the strength-to-weight ratio, size, and flexibility of the cable. The fiber arrangement and support material influence significantly the optical performance. These topics are discussed in some detail in this chapter.

Strength Members[1]

Strength members increase the permissible load a cable can withstand. If the strength member has a high Young's modulus, a high elastic limit, good flexibility, and a low weight per unit length, the result will be a fiber cable possessing the highest strength for a given cross-sectional dimension.

Some main types of material for the construction of strength members on account of their high Young's moduli are steel wires, plastic monofilaments, multiple-textile fibers, glass fibers, and carbon fibers. Some significant features of these materials are summarized in the following list*:

1. Steel wires. These have been widely used in conventional cables for armoring and longitudinal reinforcement. Various grades are available with tensile strengths to break, ranging from 540 to nearly 3100 MN/m^2. All have the same Young's modulus (19.3×10^4 MN/m^2), and the choice is guided by the preference for a high strain and yield compatible with the cable design. The main disadvantage of steel is its high specific gravity, which adds substantially to the cable weight. Furthermore, it cannot be used if a nonmetallic cable structure is required.

2. Plastic monofilaments. These are available commercially in several basic materials. Specially processed polyester filament, which combines a high elasticity modulus (1.6×10^4 MN/m^2) with good dimensional stability at elevated temperatures and a smooth cylindrical surface, is available.

3. Textile fibers. Commercial forms normally consist of assemblies of many small-diameter fibers laid up in twisted or parallel configurations. Typical examples in conventional cables are polyamides (nylon) and polyethylene terephthalate (Terylene, Dacron, etc.) with elasticity moduli which may be as high as 1.5×10^4 MN/m^2 for

---

* This description and Table 4-1 are abstracted from a paper by Baskett and Foord and are reprinted by permission.

the individual fibers. Because of the loose packing of these individual fibers, they are resilient in a transverse direction and are useful as cable fillers and binders as well as providing improved tensile properties in optical fiber cables. An exceptional member in this class which has been widely employed in optical cables is Kevlar, an aromatic polyester. The individual fibers have an exceptionally high modulus (for an organic material) of up to 13×10^4 MN/m² which, coupled with its specific gravity of 1.45, gives this fiber an effective strength-to-weight ratio nearly four times that of steel. Commercial forms of Kevlar suitable for cable reinforcement consist of composites of large numbers of single filaments assembled by twisting, stranding, plaiting, and so on, and/or resin bonding.

4. Glass fibers. For some applications, the optical fiber waveguides may supply sufficient tensile strength by themselves. Additional nonactive fibers can be used if higher strength is required. The elasticity modulus is high, typically 9×10^4 MN/m².

5. Carbon fibers. This material has been successfully employed in rigid and semirigid plastic or metal composites and has a modulus of up to 20×10^4 MN/m² in single filaments.

Relevant properties of these materials, with the exception of carbon fiber, are summarized in Table 4-1.

Besides the strength, the weight, and the elongation limit, the expansion coefficients and the cost of the strength members are equally important. The strength, the weight, and the elongation limit govern the cable size necessary to meet the strength specification, while the expansion coefficient influences the cable structural design since a high expansion coefficient relative to the fiber could cause significant fiber distortion within the cable structure over a temperature range while a low expansion coefficient could help to prevent fiber distortion. Cost is, of course, a key factor in optimizing a cable design for both performance and economy.

Cable Structures

Within a cable structure, fibers may be placed in a void region such that they experience little or no forces, or they may be placed in a fully supported fashion such that they experience only distributed hydrostatic lateral forces. In either case the intention is to prevent the optical loss and dispersion to be altered by the forces introduced by the cable structure or by externally applied forces. The external forces in the longitudinal direction are encountered during or after fiber cable deployment, for example, as the fiber cable is being pulled into a duct

Table 4-1 **Properties of Strength Member Materials**

Materials	Relative Cost	Specific Gravity	Young's Modulus, MN/m²	Tensile Strength, MN/m²	Strain at Break, %	Normalized Modulus-to-Weight Ratio	Expansion Coefficient
Steel wire	Low	7.86	19.3×10^4	$5-30 \times 10^2$	2-25	1.0	1.2×10^{-5}
Polyester monofilament	Low	1.38	$1.4-1.6 \times 10^4$	$7-9 \times 10^2$	6-15	0.3	1.3×10^{-4}†
Nylon yarn	Low	1.14	$0.4-0.8 \times 10^4$	$5-7 \times 10^2$	20-50	0.3	7.2×10^{-5}
Terylene yarn	Low	1.38	$1.2-1.5 \times 10^4$	$5-7 \times 10^2$	15-30	0.3	1.4×10^{-5}
Kevlar-49 fiber	High	1.45	13×10^4	30×10^2	2	3.5	-1.1×10^{-6}*
Kevlar-29 fiber	High	1.44	6×10^4	30×10^2	4	1.6	-1.1×10^{-6}*
S-glass fiber	High	2.48	9×10^4	30×10^2	3	1.4	1.9×10^{-6}

† Temperature range from +5 to 150°C.
* Temperature range from 0 to 100°C.

or with the fiber cable suspended between poles. The longitudinal tensile forces could cause mechanical failure of the fiber, but they introduce no modification to the loss or dispersion characteristics. The forces in the lateral direction are encountered from bending, side pressure, and so on. They introduce possible microbending, thus causing an increase in fiber loss and a modification in the multimode dispersion characteristics. These forces are externally applied or are due to the differential expansion coefficients of the cable structure to that of the fiber. The following example illustrates this situation. The fiber is placed within the cable structure such that the length of the fiber matches that of the cable structure. As temperature varies, the fiber will become either taut or slack. Longitudinal force is generated when the fiber is taut. At the same time the cable structure will be compressed. This could cause the structure to develop a kink which bends the fiber. When the fiber is slack, it needs more room. It will assume a helical position. When several fibers are loosely laid in a void space, they could be mutually entangled and crisscross each other. Severe microbending could result. When the fibers are fully supported, the compressive forces would force the fiber to assume a helical position. If the supporting material is not homogeneous, microbend could also develop.

Many cable structures have been developed on both loose and tight configurations. In general, the loose-configuration designs[2-4] result in large cross-sectional area cables suitable for use in relatively benign environments, while the tight-configuration designs result in small cross-sectional area cables suitable for use in relatively hostile environments.

Some optical fiber cable structures are shown in Figs. 4-1 to 4-3. These cables have been designed to meet specific operational requirements. The principal considerations to designing a fiber cable are summarized as follows:

- Maximum strain allowed on the fibers during the cable fabrication process, installation, and service. This determines the minimum strength of the fiber to be used in the cable structure and the amount of strength member required.
- Maximum static and dynamic lateral forces exerted on the fiber. This determines the packaging configuration and the microbend tolerance limit of the fiber.
- Adequate flexibility. This requires the fibers to be laid in a helical lay. It is to be noted that fibers are elastic and hence must be laid with uniform tension in order to achieve lay uniformity.
- Temperature range and environment over which the cable operates. This determines the type of materials to be used, particularly in terms

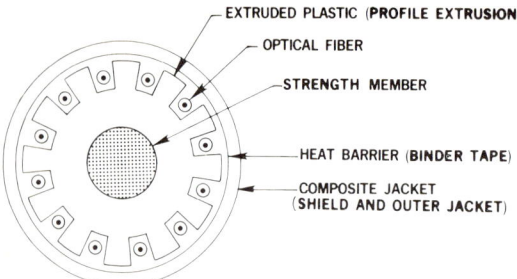

Fig. 4-1 Generic fiber cable with loose fibers in slots.

of their thermal expansion coefficient and dimensional change with moisture content.

Cable Testing[5]

The optical characteristics of the cable are evaluated by using the same measurement methods as those used for fibers. These measurements are required to be carried out while the cable is subjected to the conditions called for by the specifications. These measurements could include a nuclear survivability test in which optical loss is measured both after a short exposure to high-dosage radiation and after a long exposure to low-dosage radiation.

The mechanical tests are concerned with survivability of fiber when the cable is subjected to stresses at various temperatures and environmental conditions. These tests are generally based on standard tests used in the copper cable industry. A list of such tests is given in Table 4-2.

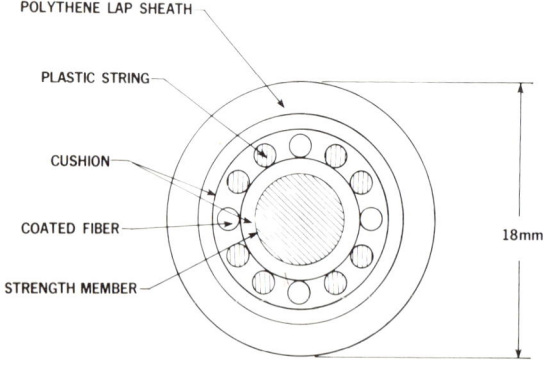

Fig. 4-2 Generic fiber cable with fiber supported by soft cushions.

Fiber Cables   81

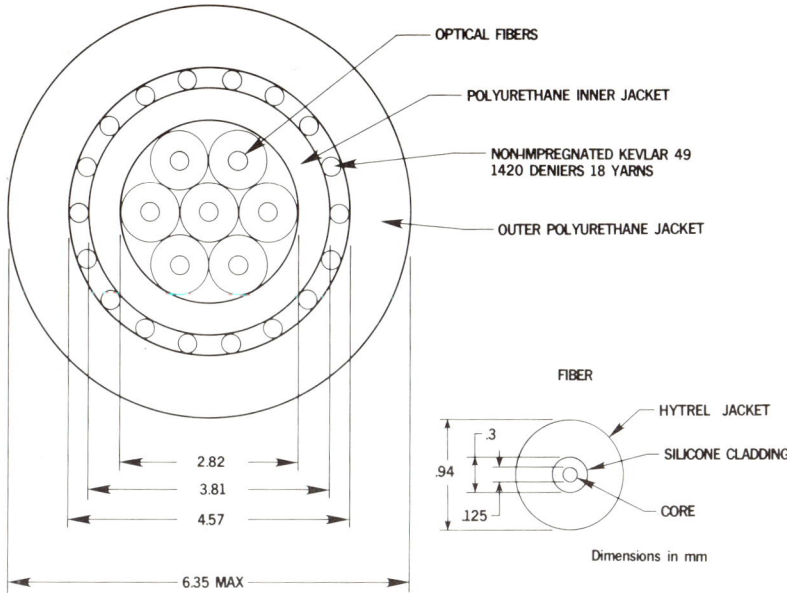

Fig. 4-3   Generic fiber cable with fully supported fibers.

Table 4-2   **Cable Testing Methods**

Tests	Examples
Mechanical evaluation	
Tensile strength test	Three samples of each type of cable are tested for fiber breakage, utilizing a 200-kg load over a 1-m gauge length; the total time to reach the required tensile strength is 26–30 s; the load is maintained for 1 min and then released in 19–28 s
Bend test	Three samples of each type of cable are tested to 2000 cycles around a diameter of 5 × cable diameter
Twist test	Three samples of each type of cable are subjected to the twist test of 2000 cycles of 360° twist over 10-cm gauge length
Impact resistance	This test is performed by dropping a 1-cm-radius spherical impact tool weighing 5 kg from a 10-cm height 200 times
Environmental evaluation	
Temperature cycling test	One sample of each type of cable is subjected to temperature cycling from −55°C to +85°C
Moisture resistance	Moisture resistance testing includes attenuation measurement after moisture cycling immersion in 98% humidity at 50°C
Fungus testing	Fungus tests are performed in accordance with Procedure 1, Method 508 of MIL-STD-810B

Cable testing is aimed at demonstrating the following:

1. *Attenuation and dispersion.* The cabling process has a negligible adverse effect on fiber attenuation and dispersion over the temperature range.
2. *Tensile test.* The cable has safely withstood tensile loading of the designed value.
3. *Impact test.* The cable has safely withstood designed impact loading at a single point on the cable.
4. *Environmental tests.* The cable has performed in a satisfactory manner at high and low temperature extremes and under conditions of vibration, moisture, and sometimes nuclear radiation.

References

1. R. E. J. Baskett and S. G. Foord, "Fiber Optic Cables," *Electric. Commun.* **52**(1):49–53 (1977).
2. C. Kao, "Optical Fiber Communication Technology," *Electric. Commun.* **54**(3):245–250 (1979).
3. M. I. Schwartz, P. F. Gagen, and M. R. Santana, "Fiber Cable Design and Characterisation," *Proc. IEEE* **68**(10):1214–1219 (1980).
4. T. Nakahara and N. Uchida, "Optical Cable Design and Characterisation in Japan," *Proc. IEEE* **68**(10):1220–1226 (1980).
5. J. C. Smith and M. Pomerantz, "Fiber Optic Cables for Local Distribution," *Proc. 25th IWCS, Cherry Hill:* 226–234 (November 16–18, 1976).

Chapter 5

Light Sources

For communication purposes, a generator of electromagnetic energy in the visible and infrared (IR) wavelength spectrum is termed a *light source*. The energy generated is the information carrier. When the light generated is at a single wavelength or frequency and maintains a uniform phase front, it is referred to as a *coherent source*. Such a light source is said to have temporal and spatial coherence and is analogous to a radio-wave oscillator, except that its frequency of oscillation is much higher. A radio-wave oscillator typically oscillates at about 10^6 Hz, while a light-wave oscillator oscillates at about 10^{14} Hz. The corresponding wavelengths are 300 m and 3 μm, respectively. Most light sources, however, generate a spectrum of light at different wavelengths. Moreover, the phase front is not uniform. Such a light source is referred to as an *incoherent light source*. It bears some resemblance to a spark oscillator which generates electromagnetic waves at a multitude of microwave wavelengths. In audio terms, the coherent source is a pure-tone generator, while the incoherent source is noise. While a change of tone represents information, the change of the loudness of noise—or more specifically, the presence and absence of noise—also can represent information.

For light sources to be suitable for application as communication signal carriers, a number of requirements are to be met. The basic requirements are power output, physical size, power efficiency, life, spectral emission modulation capability, and operating temperature range. Other requirements are coherence and cost. For fiber application, an additional basic requirement is the fiber compatibility. (Prior to the invention of lasers and high-efficiency light sources, very few light

Types of Light Sources

INCANDESCENT LAMPS

Incandescent lamps are hot-body radiators which emit a broad spectrum of electromagnetic energy according to the laws of blackbody radiation. A typical spectrum corresponding to a hot-body temperature (see the work by Feynman et al.[1]) of 4000°C is as shown in Fig. 5-1.

Physical limitations of material set the upper temperature limit for reasonably long life ($\sim$1000 h) operation at about 4000°C. The total power output is a function of the size of the filament. Since the emission spectrum is wide, extending from the visible to the infrared wavelengths, the output within a narrow spectral range is low. For communication purposes, the incandescent lamps are totally inadequate.

ARC LAMPS

Arc lamps emit light from a plasma region formed by a high-voltage breakdown of the gas at the vicinity of the electrode tips. An electric current conducted through the plasma maintains the arc. The emission

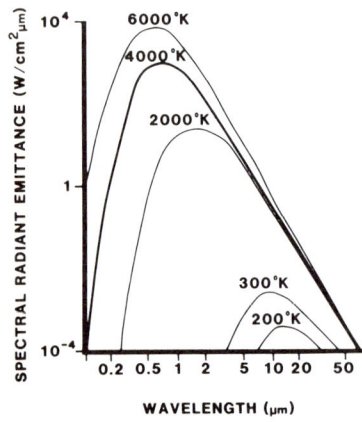

Fig. 5-1 Spectral radiations from a hot body.

is at discrete wavelengths corresponding to the electronic energy states of the gas involved.

Small arc lamps can be designed such that the lamp can be switched on and off in 1 ns. In fact, early optical communication research included work on high-speed switchable hollow cathode arc lamps using a xenon discharge. However, the light efficiency and the physical size are not ideal.

LASERS

The term "laser" is an acronym for light amplification by stimulated emission of radiation. It describes the process of light emission and amplification through the process of stimulated emission.[2] This process was first enunciated by Einstein in the laws of radiation.[3] When energy is added to an atomic structure, the electron absorbs the energy and moves to higher energy states. When it drops back to its original energy state, a radiative transition takes place, and a photon is emitted. There is a time constant involved for the decay process. This process results in random emission of radiation and is called *spontaneous emission*. Under special circumstances the electrons at a higher energy state could be trapped in that state for a time longer than the time constant of the decay process. Under this condition the population of electrons in the higher energy state builds up, and a single transition would cause an additional transition to take place in such a way that the two emissions are additive. This could induce more transitions, resulting in the emission of an amplified version of the first emission. This is referred to as *stimulated emission*. If the emitted photons are made to travel again through the region where stimulated emission can take place (the active region), the amplification will continue until the gain balances the losses and a sustained output results. The emitted radiation is coherent in time and space. In fact, the amplifier has been transformed through feedback into an oscillator. This is a brief description of a laser. The word "laser" is now used to describe a light source whose output involves the process of stimulated emission.

Lasers have been realized in many forms.[4] Most are bulky devices requiring complex drive mechanisms and have a low overall efficiency. Some lasers emit radiation with a high degree of coherence, and others can give a very large power in a narrow spectral range. For special communication purposes, such as very high capacity systems, some lasers such as the Nd YAG laser, dye lasers, and the HeNe laser can be used with advantage. The semiconductor lasers are the ones with characteristics approaching the ideal for fiber systems. The GaAlAs and the InGaAsP lasers are important examples.

LED

When a semiconductor device is constructed such that only spontaneous emission takes place, the device is referred to as light-emitting diode (LED). This device emits radiation within a relatively narrow spectral region and with sufficient output power and power conversion efficiency to be useful for optical fiber systems. Light-emitting diodes are important light sources for fiber systems.

Basic Characteristics of Light Sources for Communications

The basic light source characteristics required for application in communications are to be discussed. These characteristics govern the achievable system performance.

POWER OUTPUT

The higher the signal power output, the more attenuation that can be tolerated before the signal power falls below a level for satisfactory detection. (This refers to the power output which can be effectively utilized.) A source for an optical fiber system should have a size and radiation polar diagram (beam divergence) compatible with efficient coupling to the fiber of specified diameter and numerical aperture (NA) and should be efficiently modulatable.

The power output, however, should not create a power density at the fiber input plane exceeding the material nonlinearity limit[5] (~ 1 MW/cm^2). In practice, an output power of 10 to 100 mW is desirable.

PHYSICAL SIZE

The source may be physically bulky and yet has an optical output spot size compatible with the fiber diameter. In such cases the size presents a physical problem only. In general, a small and compact physical size makes the light source easier and more convenient to handle and gives rise to no weight or space problems.

POWER EFFICIENCY

Power efficiency determines the input power and heat-dissipation requirements. Poor efficiency means more electrical input power required for a given optical power output. This can present a power supply problem to remotely located equipment. Poor efficiency also results in large heat dissipation which must be removed by the provision of appropriate heat-removal arrangements. These aspects often increase the system cost and decrease system reliability significantly.

LIFE

When a device is used in a system in large quantities, the absolute life and the mean time before failure (MTBF) are important parameters, as they directly influence the reliability of the system.

In a stand-alone equipment, with servicing and maintenance schedules, MTBF in the order of 10^4 h is often adequate. For complex systems where many light sources are involved, it is not unusual to demand MTBF in the order of 10^6 h. However, for most of the applications, acceptable absolute life for a light source is around 10^5 h.

SPECTRAL EMISSION

The emission characteristics of a light source are required to match the designed spectral characteristics of the optical fiber transmission waveguide. The spectral loss curve of silica glass fibers shown in Fig. 2-2 indicates decreasing loss toward longer wavelengths except at OH^- absorption peaks. Regions from 0.8 to 1.6 μm have attractive low-loss transmission favorable to meeting different system requirements. At the shorter-wavelength end (0.8 μm) the loss is sufficiently low for various system applications. The longer spectral region (1.2 to 1.6 μm) is more suitable for applications where maximum distance is to be covered between repeaters. Since spectral loss varies slowly with wavelength except at absorption peaks, the loss can be regarded as constant for a source with a spectral width of a few hundred angstroms. However, if the spectral width is larger, the differential attenuation of the fiber will cause a filtering effect to occur, such that the received spectrum after transmission is narrower than the spectrum at the input end.

The source spectral width affects fiber waveguide dispersion, since the refractive index of the material varies with wavelength. This gives rise to an absolute delay time variation known as *material dispersion* and also modifies the fiber profile in the case of graded-index fibers to produce a bandwidth change known as the *profile dispersion*. At a few hundred angstroms of spectral line width, the material dispersion effect is of the same order of magnitude as the dispersion produced as a result of the difference in modal velocities in a graded-index fiber.

A further aspect relates to the possibility of using a fiber to transmit signals carried by more than one optical carrier (referred to as *wavelength multiplexing*). In this case the width of the spectral emission of the source governs the ease of achieving a high density of multiplexing. For wavelength multiplexing, the line width of emission and its shape directly influence the crosstalk levels. A set of sources with a narrow emission spectrum of a few tens of a nanometer forming a set of nonoverlapping spectral sources can be used effectively to achieve wavelength multiplexing within a spectral range. If these sources have wider emission spectrum than the overlapping criterion would allow, selective filters can be used. Care should be exercised to ascertain whether a source has a broad emission at a low power level together with the narrow emission at a high power level. The low-level radiation can significantly interfere with signals in other channels.

MODULATION

Information is imposed on the optical carrier through the process of modulation. Since the optical sources are seldom a coherent source as in radio frequency (RF) sources, phase and frequency modulation of the carrier wave is rarely used. Intensity modulation in analog as well as digital form is readily implemented and extensively used.

Few light sources can be modulated directly by varying the energy driving them. Either the driving energy cannot be varied rapidly, or the source output does not respond rapidly with the changes in drive levels. Such sources can be used only as optical carrier generators, and an external modulator must be used to execute the modulation.[6] The semiconductor lasers and LEDs, however, can be modulated directly by changing the drive current.

The rate of modulation is limited by the speed of the drive circuit and the response time constants of the light sources. A fast response enables a wider-bandwidth signal to be handled. Apart from the modulation rate, linearity is important. In fact, for analog signals, linearity is usually the limiting factor of the suitability of the light source for handling the signal. For high-quality video signals, a harmonic margin of 40 dB for second harmonic and 50 dB for third harmonic is typically required. For digital signals, linearity is much less important.

OPERATING TEMPERATURE

The ambient temperature commonly encountered ranges from a moderate range of −20°C to +40°C, to a more demanding range of −45°C to +85°C, and to an even more stringent range of −55°C to +125°C. These ranges are encountered in different terrestial applications. Even more extreme ranges may be encountered in space application.

The operation of light sources can be very sensitive to temperature. Suitable precautions such as heat sinking, heat shielding, and cooling may be required for the maintenance of operation of certain light sources over a given temperature range. For the semiconductor light sources, such precautions are definitely necessary. Semiconductor light sources not only emit less light at high temperatures, but may have a shortened life. Furthermore, temperature change causes a shift in the emission wavelength.

COHERENCE

An incoherent source resembles a noise source. It cannot be modulated by techniques such as shifting either the emission wavelength (corresponding to frequency modulation) or the phase of the emitted radiation. A coherent source resembles a radiowave oscillator and can be frequency- or phase-modulated. However, the spectral line width of a highly coherent optical source is unlikely to be less than a few megahertz, thus rendering frequency modulation with deviation of a few megahertz ineffective. Phase modulation is possible only on a comparative basis if a reference signal can be established.

The coherence effect, however, is important in systems where multipath propagation is possible. Under this condition fading of varying severity can take place, resulting in a noise component function referred to as *laser noise*. This effect is caused by small physical changes of the propagation condition as well as the shift in the spatial and temporal output of the light source caused by the modulation process. The noise generated resembles white noise and is signal-dependent.

COST

The relative cost of the source to the transmission medium is critical in the sense that a system should not be dominated by the source cost. For a system with many short links, the source cost must be kept low in order to reduce the total cost.

Source cost is composed of the actual device cost plus the cost of the associated drive equipment or circuit. Another cost is associated with the life of the device and its replacement cost. Obviously, the cost of maintenance is also important.

Selection of a Light Source

For different fibers, the light spot required for efficient launching is different. A graded-index fiber has a light-intensity distribution assuming a gaussian shape and, therefore, can be coupled with highest efficiency from a light source with the same intensity distribution. A single mode fiber has a core diameter of only several micrometers. If a source has an emission region greater than that, some light will be wasted. Furthermore, the light-acceptance angle is limited by the numerical aperture of the fiber. Thus light sources with broad radiation patterns, that is, with large divergence, will not couple efficiently.

From the discussions of the various basic light source characteristics affecting the performance of optical fiber systems, it is not difficult to anticipate that semiconductor light sources are the preferred choice, except for special applications where high power, good coherence, beam shape suitable for external modulators, and special selection of emission wavelength may favor another type of light source.

The rest of this chapter is devoted to the detailed discussion of the GaAlAs semiconductor laser and the LED and a brief discussion of the InGaAsP system to illustrate its characteristics.

PHOTOEMISSION FROM SEMICONDUCTOR DEVICES[7]

Photoemission may take place in the semiconductor *p-n* junction device when electrons and holes are injected, respectively, into the *p-* and *n*-type regions under the influence of a voltage applied in the forward direction, that is, to decrease the voltage at the junction. The injected carriers are referred to as *minority carriers*. Under *forward bias,* as this charge flow injection condition is called, the minority carriers, as opposed to the majority carriers caused by dopants in the semiconductors, will recombine, mainly with the majority carriers, as they travel near the junction. They recombine either radiatively or nonradiatively. In a radiative recombination the injected minority carrier looses a certain amount of energy. This energy appears as a photon energy h_v. This energy is approximately controlled by the potential of the *p-n* junction, which is known as the *bandgap*. Photoemission has taken place. On the other hand, if the energy loss is converted to heat, the process is a nonradiative recombination. The selection of the processes depends on the lifetimes of the minority carriers. The energy band diagram of a *p-n* junction for photoemission is illustrated in Fig. 5-2.

The lifetime of a recombination process is controlled by the energy-band structure and the density of nonradiative recombination centers and their capture cross sections. The nonradiative recombination cen-

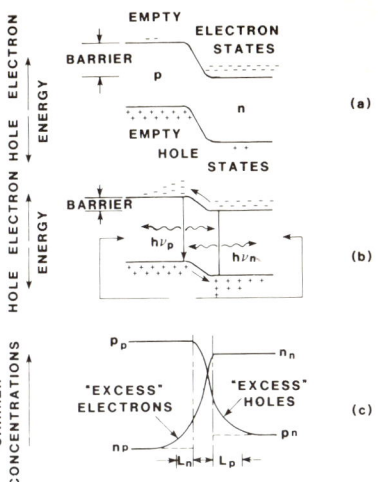

Fig. 5-2 Electroluminescent *p-n* junction operation. *(a)* Zero bias. The built-in potential drop across the *p-n* junction represents a large barrier for the motion of the electrons and holes. *(b)* Forward bias. The potential barrier is significantly reduced by the application of the external voltage. *(c)* Majority and minority carrier concentrations on the *n* and *p* sides of a forward-biased *p-n* junction.

ters are usually associated with material imperfections where the kinetic energy of the carriers converts into heat. The internal efficiency η_i and the lifetime τ, are defined by

$$\eta_i = \frac{1}{1 + \tau_r/\tau_{nr}}$$

$$1/\tau = \frac{1}{\tau_r} + \frac{1}{\tau_{nr}}$$

where τ_r and τ_{nr} are the radiative and nonradiative lifetimes, respectively.

For application as a light source, efficient radiative transitions are important, while nonradiative transitions should be minimized. Furthermore, for high-speed operation, the recombination lifetime should be as short as possible. These criteria are used in selecting materials for light sources. In general, semiconductors with a bandgap across which the electrons and holes have the same momentum make recombination possible without involving a third particle. This type of recombination, called *direct transition,* is efficient and has short recombination lifetimes. Comparing GaAs, a direct-bandgap semiconductor, with indirect-bandgap semiconductors such as silicon, the calculated radiation lifetime for silicon at majority carrier density of $10^{17}/cm^3$ is about 10^{-4} s, which is four orders of magnitude longer than that for GaAs, which is 10^{-8} s. Two important direct-bandgap semiconductors are GaAlAs and InGaAsP, for which the discussion is concentrated. A range of semiconductors involving two (binary), three (ternary), or four (quaternary) elements is shown in Fig. 5-3. The lattice spacing of these materi-

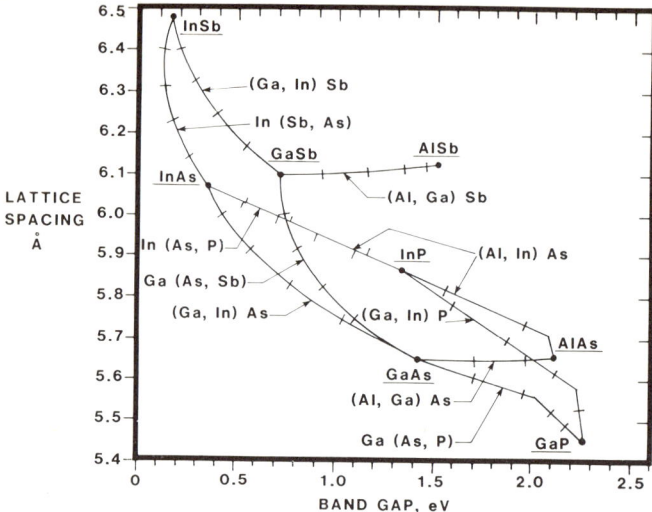

Fig. 5-3 Lattice spacing and bandgap of regions III to V compound binary, ternary, and quaternary semiconductors.

als of different composition is plotted against bandgap energy, thus allowing the important aspect of lattice match (to be discussed later) to be established in a device as well as providing a convenient chart for locating bandgap energies of various semiconductors for emission wavelength selection purposes.

When the injected carriers create a population distribution such that an energy state with a higher energy has a population larger than that in the lower-energy state, inverted population is established. In this situation radiative transition of a photon with energy equal to the difference of energy between these two energy states can be induced by a photon of that energy. In other words, if a photon of this energy happens to be in the region where the inverted population exists, it will have a smaller probability of being absorbed than inducing a radiative transition. The induced radiation, moreover, will be of the same photon energy and at the matching phase, and the resulting radiation will be coherent and amplified. This form of emission is referred to as *stimulated emission*. If the photons built up in one direction of travel are dominant, and if the gain is higher than the loss, a sustained amplification can take place, resulting in a directional output. A laser is a coherent emitter in which the photons are made to travel repeatedly through the amplification region by providing at each end of the region a reflecting surface. The width of the laser spectral output is governed by the cavity Q.

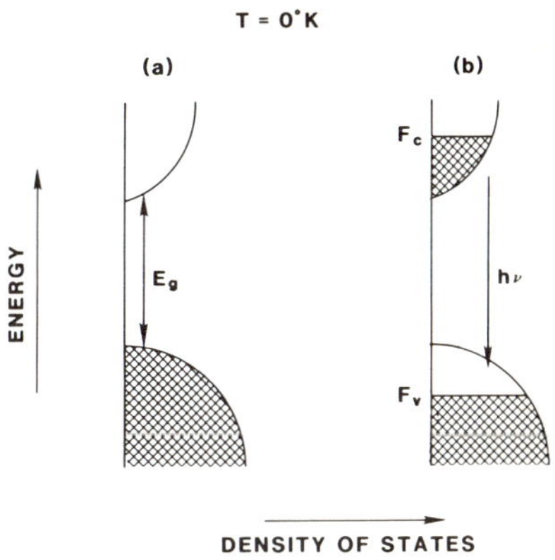

Fig. 5-4 Electron energy as a function of the density of states in an intrinsic direct bandgap semiconductor at $T = 0$ K in equilibrium (a) and under high injection (b).

In a semiconductor laser the energy levels of the carriers are not distinct.[8] The energy levels are more appropriately described by energy bands. The inverted population when high injection takes place is represented in an energy-momentum diagram as shown in Fig. 5-4. Emission will take place at $h\nu$ corresponding to $F_c - F_v$, where F_c and F_v are the energy levels commonly referred to as *quasi-Fermi* levels. At lower injection stimulated emission is not sustained; therefore, only spontaneous emission will occur. The spontaneous emission spreads over the entire energy gaps and will have a peak where the competing losses are at a minimum. The stimulated emission will occur near this loss minimum. The threshold injection level for laser action is called the *lasing threshold*, and the corresponding current is called the *threshold current*.

The actual energy levels involved in a semiconductor laser with highly doped materials are complex. The radiation usually corresponds to an energy less than that of the bandgap.

In order to make an efficient laser, the physical structure is designed to provide means to confine the carrier and the radiated optical energy such that a controlled amount of injected current can result in the desired emission efficiency, output power, radiation pattern, threshold current stability, and output power/current linearity.

GaAlAs LASER STRUCTURE

A GaAlAs laser is a semiconductor laser[9] emitting in the wavelength region from about 0.8 to over 0.9 μm. Gallium arsenide is a direct-bandgap material with favorable material characteristics for high-speed light source applications. The addition of aluminum alters the bandgap and expansion coefficient but not the lattice spacing, so that highly

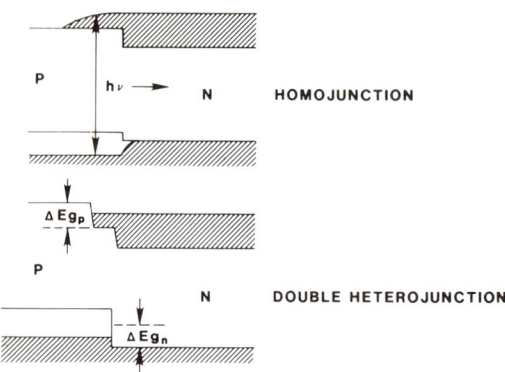

Fig. 5-5 Energy-band diagrams of a *p-n* junction.

efficient devices capable of continuous-wave (CW) operation at elevated temperatures and with long operating life can be created.

The injected carrier density needed to reach threshold for laser action is about $10^{18}/cm^3$. To achieve this density across a *p-n* junction, it is necessary to prevent the carriers from freely diffusing across the junction. Suitable potential barriers for minority carriers can be created in the regions adjacent to the junction at distances smaller than the diffusion zone. This is illustrated in Fig. 5-5. It can be seen that an effective carrier confinement is achieved by introducing a pair of potential barriers in the *p* and *n* regions. Each potential barrier is said to have created a heterojunction; hence the new device is called a *double-heterojunction device*. It requires a lower current to achieve a given current density. This means that the lasing threshold can be reached with a lower injection current. The electrical-to-optical conversion for stimulated emission efficiency is also improved.

Within the region of high carrier concentration, stimulated emission takes place. If two parallel ends of this region are created, the emitted radiation will resonate within the cavity so created. The width between the two heterojunctions will be the width of the emission zone. It turns out that the carrier concentration increases the refractive indices of the junction. Furthermore, the material refractive indices outside the heterojunctions are lower than those within the region. As a result, the region near the junction acts as an optical waveguide, providing guidance to the radiation from the stimulated emission. In principle, it is possible to optimize the carrier confinement region and light confinement region independently by further adjustment of the potential barriers and material indices. This sophistication is to be traded with fabrication difficulties.

The area of the junction has not been considered thus far. The argu-

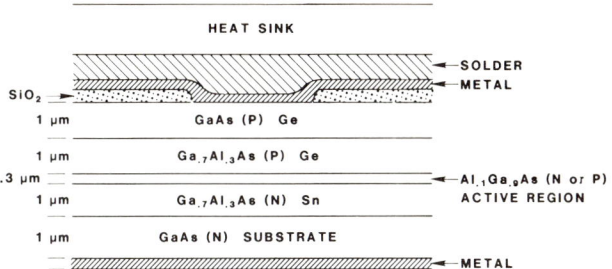

Fig. 5-6 Double-heterojunction stripe geometry laser.

ment given above applies to a device of arbitrary area. The width and the length of the active region are important structural design parameters. The length-to-width ratio affects the device efficiency. The width also controls the radiation pattern and the number of permitted lateral emission mode patterns. For a fiber compatible device, the width should be less than or equal to the fiber core diameter. Furthermore, threshold current is directly related to the width and the length.

A practical device in the GaAlAs system incorporating the above features can be feasibly fabricated by using GaAlAs materials of different bandgaps.

A successful structure[10] is as shown in Fig. 5-6. This incorporates an oxide isolated stripe contact to define the width of the emission region and a double-heterojunction structure created on planar geometry. This structure is the simplest approach to achieve the desired geometry and is used here to illustrate the desired principle.

The overall size of the device is around 200×500 μm^2 with a stripe width of several micrometers. The emission region has a width of around 20 μm. Such a device has a threshold current of 150 mA and a corresponding threshold current density of 1000 A/cm^2. The emitted power is about 10 mW and can operate as a CW source up to a maximum temperature of at least 70°C. Its life is in the order of 10^5 h. The emitted radiation has a single-mode pattern and generally oscillates at many longitudinal cavity resonant frequencies; hence the output has a spectral width of several tens of angstroms.

Other structures[11] can provide current and optical confinement and result in single transverse mode operation at low threshold currents.

GaAlAs LASER CHARACTERISTICS

A GaAlAs laser device for optical fiber application is designed to achieve specific electrical and optical characteristics.[12]

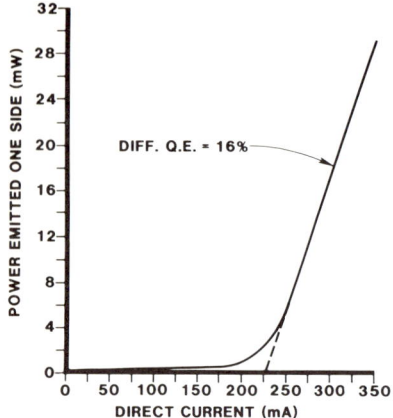

Fig. 5-7 Laser output characteristics showing the lasing threshold.

Typical power output versus current characteristics are as shown in Fig. 5-7. The laser is designed to give good linearity by maintaining a constant differential quantum efficiency to the maximum power output point. It is important to note that because of the finite thermal conductivity between the junction and the heat sink and the diode series resistance (typically 0.5 Ω), the junction temperature is higher than that of the ambient. Laser emission efficiency and wavelength are, therefore, dependent on the junction temperature. The wavelength variation is about 0.1 Å/°C. This means that both the output power and the emission wavelength of the laser will vary with temperature. Furthermore, the variations have a signal dependence as well as a dependence on ambient temperature change. The latter can be corrected by an average power-based feedback control, while the former can be corrected only by a fast-feedback circuit. Usually, the output from the laser is sufficiently stabilized by a slow-feedback circuit which adjusts the bias current to a fixed value just above the threshold current and which often has a maximum power output control to limit the maximum drive current so that within the permitted range of current, the laser output does not exceed the power rating of the device for linear operation. Under this set of operation conditions the laser output power varies linearly with drive current and has a high quantum efficiency. The maximum modulation speed is governed by the time constants of the recombination process under the stimulated emission condition. The recombination constant is faster than 10^{-10} s. Thus, with a suitable drive circuit, lasers can be modulated at a speed in the gigahertz region. However, at such extremely high frequencies the behavior of the laser is complex and is not well defined.

OPTICAL CHARACTERISTICS

The active region of the GaAlAs laser diode is an optical waveguide of rectangular cross section. The height of the rectangle is a fraction of a micrometer, while the width for a narrow-stripe laser designed for single transverse-mode operation is several micrometers. The refractive index difference between the active region and its vicinity is such that the device can operate with a single transverse-mode pattern. If the width is made larger, then more than one transverse mode can exist. However, even under this condition the device emits in a single mode when the drive current is just over threshold. This is because the emission is associated with a filamentary zone of the highest gain where the refractive index is also the highest. Thus a laser device with a wide emission stripe tends to have a filamentary emission zone across the stripe and, as drive current increases, to have more than one zone across the stripe. Statistical variations often cause this situation to be not always repeatable. Filamentation causes coupled power to a fiber to vary as well as nonlinearities in the output power versus drive current characteristics. A single transverse-mode structure is preferred for a device designed for optical fiber applications.

For a single transverse-mode device, the emission wavelength λ is controlled by the bandgap of the material as well as the allowed longitudinal modes defined by the parallel-plate Fabry-Pérot cavity. The mode spacing for a cavity of length L in a region refractive with n is

$$\frac{\lambda^2}{2L(n - \lambda\, dn/d\lambda)}$$

For a typical laser, this corresponds to 2 to 3 Å. The spectral width of a laser depends on the number of such modes present at a particular drive condition. While it is usual to have several longitudinal modes present, it is possible to achieve a single longitudinal-mode operation by reducing the length of the waveguide cavity. This effectively reduces the gain of adjacent emission lines to the extent that only one line is favored.

In a device operating in a single transverse mode the near-field intensity distribution is very narrow in the direction normal to the junction and much wider in the transverse direction. The intensity distribution in the transverse direction is approximately gaussian. Such a near-field distribution gives rise to a radiation lobe which has an elevation half angle of 30 to 50° and an azimuthal half angle of about 10°.

The Fabry-Pérot cavity of the laser is formed by cleaving the GaAs substrate at appropriate crystal planes, thus ensuring a high degree of parallelism. To enhance the cavity Q, one facet is usually coated with a reflecting material to increase the reflectance which ordinarily arises from the refractive-index difference between GaAlAs ($n \sim 4$) material and air. The emission is mainly from one facet. This facet is usually coated with a passivation layer to avoid facet damage, which can result in device failures. The back facet, although coated, still emits energy sufficient for use in power-monitoring purposes.

The emission from a single longitudinal mode of the laser is coherent. The degree of coherence can be expressed in terms of the length of the emitted wave train along which the relative phase of the front portion to the tail portion can be unmistakably correlated. The coherent length of the GaAlAs laser was found to increase with the increase in drive current. When a number of longitudinal modes are present, the apparent coherence is decreased. The coherent nature of the emission, however, gives rise to interference when multipath propagation is possible. It is to be noted that a low-level radiation of spontaneous emission is also present along with the stimulated emission.

GaAlAs LIGHT-EMITTING DIODES

If the GaAlAs laser is not equipped with the Fabry-Pérot cavity, stimulated emission can be suppressed in favor of spontaneous emission. By suitably tailoring the device structure, light-emitting devices known as *light-emitting diodes* (LEDs) can be fabricated.

GaAlAs LED Structure

The simplest LED structure is a *p-n* junction. However, the light emitted at the junction is not readily accessible. Light-emitting diodes suitable for use with the optical fiber can be constructed by tailoring the structure such that emission is from a small region and internal reabsorption is minimized and an output region defined.

Two structures have been successfully developed.[13-15] Both employ a heterojunction to allow the buildup of a high concentration of carriers for efficient recombinations. One is based on the laser structure, except that the emission region is confined to a short portion of the cavity length in order to reduce internal absorption and hence increase the overall efficiency. The emission is from the edge of the device. This is shown in Fig. 5-8. The radiation is partially directional as a result of the waveguide effect. The emission, however, is incoherent, and the radiation lobe has a half angle of about 60°. The facet surface of the LED is usually coated with an antireflection coating, and the emission region width is arranged to match the core diameter of the fiber.

Light Sources 99

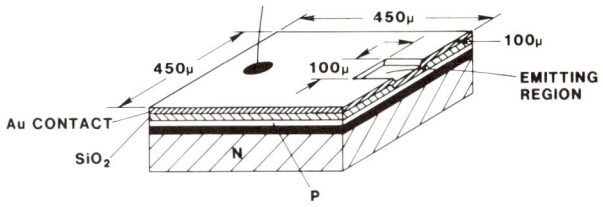

Fig. 5-8 Edge-emitting LED.

A surface-emitting structure is as shown in Fig. 5-9. The current flow is confined by the use of a small circular metallic contact through a silica mask. The current entering the *p-n* heterojunction can cause a region (slightly larger than the contact) to have a high concentration of carriers. Recombination takes place, and light is spontaneously emitted. An etched pit is created opposite to the contact, thus permitting the emitted radiation to reach the outside world with little attenuation. The etched pit allows a fiber to be inserted to collect the omnidirectionally radiated light with maximum efficiency.

Electrical and Optimal Characteristics

A surface-emitting diode operates at a current density in excess of 1000 A/cm² and emits about 1 to 10 mW. The emission wavelength

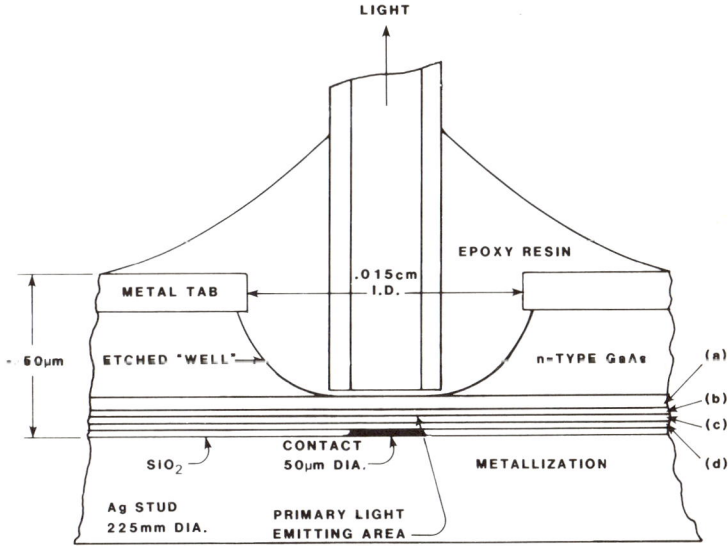

Fig. 5-9 Cross section of a surface-emitting LED: *(a)* n-type AlGaAs; *(b)* p-type AlGaAs (active layer); *(c)* p-type AlGaAs; *(d)* p-type GaAs.

is centered around the bandgap of the material system, and the emission line width for GaAlAs extends over about 300 Å. The radiation from the device is omnidirectional, and typically the radiation coupled into the fiber is about 100 µW. The contact is about 125 µm in diameter, and the actual current needed to drive the device is around 300 mA.

An edge-emitting diode emits marginally less power than a surface-emitting diode but can couple to a fiber somewhat more efficiently. The drive current requirement is also similar.

The response time is governed by the recombination time constant. With spontaneous radiation, the recombination constants are much longer than those involved in a stimulated radiation process; it is in the order of 10^{-8} s. The modulation speed limit is related to the time constant as follows:

$$\frac{P(w)}{P_0} = \frac{1}{[1+(w\tau)^2]^{1/2}}$$

where τ is the minority carrier lifetime, P_0 is the emitted power at zero frquency, and $P(w)$ is the emitted power at frequency w.

This means that τ must be as small as possible. This can be achieved by increasing the dopant concentration. However, at high doping levels a high density of nonradiative centers are formed. These will increase τ. By trading off these two factors, a modulation speed of 100 MHz can be achieved.

Reliability

The life of a semiconductor GaAlAs light-emitting device has been found to depend critically on the material perfection.[16] Imperfections lead to current concentrations, local temperature increases, and other parameters, which induce the growth of such regions to larger sizes and decrease emission efficiency, thus eventually causing the device to fail. In a laser the imperfection within the gain region has a magnified effect on the emission characteristics. Hence lasers call for more stringent control on the fabrication and mounting processes in order to minimize the occurrence of imperfections and stress centers, which have been shown to induce movement of imperfections.

In practice, the temperature of operation, the magnitude of the current density, and the power output level can affect the life of the device. The higher the temperature, current density and output, the faster the rate of degradation. The rate dependence is nonlinear. Thus a 10^5-h-life device at room temperature may have only a 10^3-h life at 70°C.

For the laser, the output power can be at a power level sufficiently high to cause material damage to the facet. This occurs at a power density level of $\sim 2 \times 10^6$ W/cm². For this reason the peak power of a laser should be limited. Facet damage can be reduced by the use of a dielectric coating which decreases the intensity of the electric field caused by the optical wave.

InGaAsP MATERIAL SYSTEM

A ternary semiconductor enables light-emitting devices to emit light at different wavelengths, since the addition of a third element alters the bandgap energy. In the case of GaAlAs, this turns out to be a successful material system for creating a range of devices for light emission at a range of wavelengths from 0.8 to 0.9 μm, and at the same time enables the structure to be fabricated with little lattice mismatch. In general, the ternary material systems offer the prospect of making devices for emission at different wavelengths, but the lattice match condition may not be fulfilled. In these cases the resulting devices are unlikely to work very efficiently, nor will they operate over a long time without failure. The introduction of a fourth element for certain semiconductor material systems enables both emission wavelengths to be controlled and allows the lattice match conditions to be simultaneously fulfilled; InGaAsP is such a material system, with an emission range of 1 to 1.8 μm.[17]

The difficulty of making an InGaAsP or any other quaternary semiconductor lies in the increased number of variables and, hence, the complication of establishing a viable fabrication technique which would enable the required solid composition to be established from the constituents in liquid or vapor states. Using InP as the substrate, InGaAsP has been successfully grown by using both liquid-phase and vapor-phase epitaxy to create a range of devices for emission at wavelengths of 1.1 to 1.8 μm and with lattice match established (see Fig. 5-3, which illustrates the emission wavelength and lattice match conditions).

The improved lattice match enables long-life devices to be made successfully. Both LEDs and lasers can be designed on the same principles as in the case of the GaAlAs devices.

The InGaAsP LEDs emit with a wider spectrum typically 600 Å, while the temperature effect[18] on the emission wavelength for both LEDs and lasers is also higher. The recombination time is short so that high-speed devices comparable to the GaAlAs devices can be achieved.

The importance of the light sources based on this material system can be seen by referring to Fig. 1-2. It can be clearly seen how these light sources could allow the lower loss transmission regions of the silicon fibers to be exploited.

References

1. R. P. Feynman, R. B. Leighton, and M. Sands, *The Feynman Lectures on Physics,* Vol. 1, Chapter 41–2, and Vol. III, Chapter 4–5.
2. A. L. Schawlow and C. H. Townes, "Infrared and Optical Masers," *Phys. Rev.* **112**:1940–1949 (1958).
3. R. P. Feynman, R. B. Leighton, and M. Sands, *The Feynman Lectures on Physics,* Vol 1, Chapter 42–5, and Vol. VII, Chapter 9–6.
4. J. E. Geusic, W. B. Bridges, and J. I. Pankove, "Coherent Optical Sources For Communications," *Proc. IEEE* **58**(10):1419–1439 (1970).
5. P. Labudde, P. Anliker, and H. P. Weber, "Transmission of Narrow Band High Power Laser Radiation Through Optical Fibers," *Opt. Commun.* (*Netherlands*) **32**(3):385–390 (1980).
6. F. S. Chen, "Modulators for Optical Communications," *Proc. IEEE* **58**(10):1440–1457 (1970).
7. A. Yariv, *Quantum Electronics,* Wiley, New York, 1967; see also H. Kressel and J. K. Butler, *Semiconductor Lasers and Heterojunction LED's,* Academic Press, New York, 1977.
8. G. H. B. Thompson, *Physics of Semiconductor Laser Devices,* Wiley, New York, 1980, Chapter 2.4.
9. M. B. Panish, "Heterostructure Injection Lasers," *Proc. IEEE* **64**(10):1512–1540 (1976).
10. J. E. Ripper, J. C. Dyment, L. A. D'Asaro, and T. L. Paoli, "Stripe-Geometry Double Heterostructure Junction Lasers: Mode Strucure and C. W. Operation Above Room Temperature," *Appl. Phys. Lett.* **18**:155–157 (1971).
11. G. H. B. Thompson, *Physics of Semiconductor Laser Devices,* Wiley, New York, 1980, Chapter 6.2.
12. A. A. Bergh and J. A. Copeland, "Optical Sources for Fiber Transmission Systems," *Proc, IEEE* **68**(10):1240–1247 (1980).
13. C. A. Barrus and B. I. Miller, "Small-Area Double Heterostructure Aluminum-Gallium-Arsenide Electroluminescent Diode Sources for Optical-Fiber Transmission Lines," *Opt. Commun.* **4**:307–309 (1971).
14. J. P. Wittke, M. Ettenberg, and H. Kressel, "High Radiance Light Emitting Diodes for Single-Fiber Optical Links," *RCA Rev.* **37**:159–183 (1976).
15. D. Marcuse, "LED Fundamentals: Comparison of Front and Edge-Emitting Diodes," *IEEE J. Quantum Electron.* **QE-13**(10):819–827 (1977).
16. M. Ettenberg and H. Kressel, "The Reliability of (Al_1Ga) as CW Laser Diodes," *IEEE J. Quantum Electron.* **QE-16**:186–196 (1980).
17. G. H. B. Thompson, *Physics of Semiconductor Laser Devices,* Wiley, New York, 1980, Chapter 3.1.
18. G. H. B. Thompson and G. D. Henshall, "Non-radiative Carrier Loss and Temperature Sensitivity of Threshold in 1.27 μm (GaIn) (AsP)/InP DH Lasers," *Elect. Lett.* **16**:42–44 (1980).

chapter

Modulation and Detection

Introduction

MODULATION

The process of placing information onto an information carrier is called *modulation,* while the converse process is referred to as *demodulation.*[1] If the carrier is a single-frequency source (sometimes described as a line source in optical spectroscopy sense), the information signal can be imposed on the carrier as an amplitude variation, frequency variation, or phase variation. In symbolic form, if $A(w)$ is the signal and $\sin w_0 t$ is the carrier then

$A(w) \sin w_0 t$ is amplitude modulation
$\sin(w_0 - \Delta w)t$ is frequency modulation
$\sin(w_0 t + \Delta \phi)$ is phase modulation

Demodulation can be achieved by the detection of the amplitude variation in the case of amplitude modulation and by the use of a circuit which produces an amplitude variation proportional to frequency or phase variations.

It is easily demonstrated that the higher the carrier frequency, the larger its potential to carry a wide band of information. It is this concept which led to the conclusion that communication channels at optical frequencies have enormous bandwidth. Of course, this is true, if the

103

carrier is a line source. Otherwise, the available bandwidth is diminished. In fact, the bandwidth offered with the use of an optical carrier is more likely to be limited by the baseband electronics and modulator-demodulator designs rather than the carrying capacity of the optical carrier.

The carriers at optical wavelengths range from relatively incoherent emitters such as LEDs to relatively coherent emitters, such as semiconductor lasers. The line widths are in the order of several hundred angstroms for the LED operating at a center wavelength of 0.85 μm. This corresponds to a frequency spread of about $\pm 10^{13}$ Hz from a center frequency of 3.5×10^{14} Hz. For a multimode laser, the spread is about $\pm 0.4 \times 10^{12}$ Hz and for a single longitudinal mode laser, plus or minus tens of megahertz. This means that direct frequency or phase modulation is inappropriate at optical wavelengths, except under exceptional circumstances when special arrangement is made to enable phase or frequency to be detected. The modulation processes which can easily be applied to an optical carrier essentially produce intensity variation of a noiselike carrier.

The signal format and the method to produce the intensity variation can be chosen to combat noise and distortion.[2] For an analog signal, it can be first amplitude-, frequency-, or phase-modulated onto a subcarrier. This often facilitates the design of filters for the separation of different channels of information carried on a FDM (frequency-division multiplex) basis. It can be used to avoid cross-modulation products, as a result of system nonlinearity. It can be arranged to improve noise performance. The subcarrier with the information intensity-modulates the optical carrier. At the receiving end the subcarrier is first intensity-demodulated, and the information on the subcarrier is extracted by demodulators. The analog signal can also be converted first to a digital form. The digital signal then modulates the optical carrier directly or first processes it into frequency-shift keying (FSK) or phase-shift keying (PSK) format. The use of FSK or PSK can offer certain discrimination to be available after demodulation and allow errors to be controlled more easily.

Methods available for the production of intensity variations of an optical carrier depend on the type of optical source. For a semiconductor source, intensity variations can be directly produced by the variation of the drive current. For linear modulation, the source is biased at a convenient point to allow an increase or a decrease of output intensity corresponding to the signal-level changes over a linear characteristics region. For pulse signals of the binary type, the device can initially be biased at a point to produce little or no output. This lowers total power consumption as well as making the receiver design simpler, by eliminating the need to deal with a steady dc current.

Alternatively, an external modulator can be used for most light sources. External modulators (see the paper by Chen, Ref. 6 of Chap. 5) utilize electrooptic, acoustooptic, magnetooptic, and band-edge shift effects to produce path length, polarization, or absorption changes. These modulators produce intensity changes directly, except for polarization modulation where polarizer-analyzer assembly is used to translate polarization changes to intensity changes.

NOISE AND DISTORTION CONSIDERATIONS

The signal is degraded by noise and distortion accompanying the process of modulation, transmission, and detection. The noise term contributed by the detection process is the most dominant, while distortions could arise at each of the three stages of modualtion, transmission, and detection. When a laser source is used, the noise associated or indirectly associated with the transmitter can also be significant.

Associated with the transmitter, the nonlinearity of the modulation characteristics and the frequency response of the source introduce harmonic distortion and an upper frequency limit. The perturbation of the output power from the source due to signal-dependent reflected power is a noise term. The signal-related spectral variations of a coherent source could give rise to another noise term known as *laser modal noise*,[3] when the propagation path is a multimode fiber with cross-sectional area mismatches. The spectral variations give rise to a change in the interference pattern and result in different amounts of power being transferred at the mismatched joints. These noise terms can be large and limit the signal-to-noise ratio (SNR) to 40 to 50 dB at the transmitter output port.

Associated with the optical fiber transmission line, the waveguide dispersion, consisting of modal, spectral, and material dispersions, is the principal distortion term. No significant noise is added to the signal during transmission, except in instances where the waveguide has discontinuities sufficient to give rise to multiple echo distortion or has appreciable discontinuities at the core-cladding interfaces to give rise to mode conversion and reconversion from core to cladding modes.

Associated with the detector, nonlinear distortion is insignificant, but the noise term is dominant. As discussed later, the thermal noise is the dominant noise for a photodiode receiver while a balance of excess noise from the multiplication process and thermal noise form the dominant noise for an avalanche photodiode (APD) receiver.

SIGNAL-TO-NOISE RATIO REQUIREMENTS

The SNR requirement for a signal varies for different applications.[4] For an analog signal, the required SNR is rather large. For a telephone signal, since our ears have a large dynamic range and the signal level can vary greatly between speakers, the noise is required to be below audibility when a soft-spoken person is speaking. For a television signal, a 42-dB SNR of peak-to-peak signal and root-mean-square (RMS) noise is generally acceptable to an audience. For a music channel, a 50-dB SNR may still offend a hi-fi connoisseur's ears. This type of SNR requirement is difficult to meet in an optical fiber system where the transmitted power is around 0 dBm and the noise level of the receiver is of the order of 6 pW/Hz; hence for a 20 kHz signal, the SNR, even with no loss present, is only 57 dB. One way to reduce the SNR requirement for an analog signal is to process the signal in frequency-modulated (FM) format on a subcarrier.

The signal is still an analog signal, but the use of frequency modulation trades sensitivity to intensity variation for sensitivity to expanded bandwidth. From classical frequency-modulation theory, the SNR improvement is related to the index of modulation, which is essentially a bandwidth-expansion factor. The improvement factor is $(\Delta f/f_m)^2$. The conversion of the signal into an FM format involves an extra modulation stage. The signal is placed on a subcarrier, and the bandwidth can be several times larger. This means that the optical intensity modulation is at a higher frequency. The highest frequency permissible is set by the frequency response of the light source and the photodetector. This sets the limit to the noise improvement attainable.

The SNR required for digital signals is much lower. In optical fiber systems the digital signal is in a binary form. The actual SNR required depends on the error rate tolerable by the system. For a 10^{-9} bit-error rate (BER), an SNR of about 20 dB is more than sufficient. This type of SNR is readily achieved in an optical fiber system.

SIGNAL DETECTION

Signal detection at the receiver is achieved by the use of photodetectors together with their associated electronic circuits. The aim is to convert the signal—which is in the form of intensity variation of the optical carrier—to its electrical form, and then to restore the signal with a minimum addition of noise and distortion to its original state through appropriate amplification. The receiver performance is governed by the choice of the type of photodetector as well as the design of the associated electronic circuit. A receiver is usually designed to minimize the noise power contributed by the receiver, in order to minimize the amount of signal power required to produce a given SNR. In other words, the receiver is designed to achieve a high sensitivity which permits a large attenuation between the transmitter and the receiver for a given transmitter power output and a given required SNR. The larger attenuation margin usually implies longer achievable transmission length between the transmitter and the receiver—an important parameter in the control of system design flexibility and system cost.

Photodetectors

Photodetectors for application in communications should have a high quantum efficiency at the appropriate spectral region, adequate frequency response, low dark current, and low signal-dependent noise. Several types of photodetectors exist[5]: the photoemissive type, the photoconductive type, and the photovoltaic type. The photovoltaic type has the best overall performance and is the most appropriate type for optical fiber system application.

PROPERTIES OF PHOTODETECTORS

Quantum Efficiency

Photodetectors convert the incoming photons to electrons. The photon to electronic conversion efficiency is known as its *quantum efficiency*. It is 100 percent efficient if for every incident photon, a photon-induced electron is released.

Spectral Response

The photodetector material absorbs photons over a spectral range. This spectral range is governed by the material electronic band structure. The incident photon must exceed the energy of the bandgap to release electrons.

Frequency Response

The construction of the photodetector determines the speed of response to an impulse or a CW signal. For resolution of the pulses, the rise and fall times of the device must be rapid. A flat bandwidth frequency response is required for handling a signal occupying a certain bandwidth.

Dark Current and Signal-Dependent Noise

In the absence of signals, the residual electron emission is called the "dark current." It is an unwanted noise. For some devices, the electrons released by the signal have a random component whose amplitude is dependent on the level of the signal. This signal-induced noise occurs in certain devices.

TYPES OF PHOTODETECTORS

Photoemissive Devices[6]

Photoemission is the emission of electrons from a photocathode when incident photons are absorbed by the cathode material. The photoemissive materials are materials with low work function (energy required to cause an electron to detach from the material lattice to become a free electron). When a photon enters the material, it collides with the material and gives up its energy to an electron within the material. The energized electron will be released if the acquired energy is greater than the work function of the material. Most material has work functions higher than the energy represented by a photon with a wavelength of 1 μm. Hence most photocathodes have no response to light at wavelength >1 μm. Some representative material and its quantum efficiency over a spectral region are as shown in Fig. 6-1.

A simple photoemissive device is a phototube in which a photocathode and an anode are placed in a vacuum glass envelope. The emitted electrons caused by the incident photon are collected by the anode when a bias voltage is applied. The vacuum is needed to allow

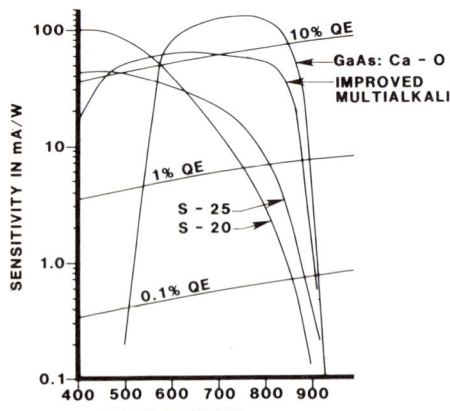

Fig. 6-1 Spectral response of photocathodes.

free passage of the electron from the cathode to the anode and to prevent the cathode from contamination.

Quantum efficiency typically is 1 to 10 percent but decreases rapidly with wavelength approaching 1 μm. Spectral response extends from about 0.4 to 0.9 μm. Frequency response can be as high as several gigahertz. Dark current depends on the type of photocathode and the operating temperature. It can be a significant noise source.

With the use of secondary emitters, the electrons emitted from the photocathode can be noiselessly amplified. The secondary emitters are called *dynodes*. These are biased at successively higher potentials so that the electrons are accelerated toward them at each stage. When an accelerated electron strikes the dynode, it would cause more than one electron to be released. This multiplication process can provide doubling or even quadrupling of the number of electrons per stage. A total gain of 10^6 can be achieved in this type of an arrangement. A phototube with dynodes is called a *photomultiplier*.

The quantum efficiency and the spectral response are the same as those of the phototube. The time response is limited by the spread of the arrival time of the emitted electrons. This gives rise to a typical bandwidth of several tens of megahertz, although a bandwidth of several hundred megahertz is possible in special tubes. Dark current can be relatively low. The photomultiplier with high gain can achieve extremely high sensitivity for relatively low frequency signals. Its major disadvantage is its bulky physical structure, complex bias arrangement, and high cost.

In general, photoemissive devices are seldom considered for optical fiber system except for special applications, such as in instrumentation.

Photoconductive Devices

In insulators such as CdS and intrinsic semiconductor materials, such as PbS, PbTe, and extrinsic semiconductors, such as doped germanium and silicon, the incident photons cause electrons to move into the conduction band, thus increasing the conductivity of the material. If a bias voltage is applied, the standing current in the circuit would alter when the bulk photoconductive material is illuminated by the incident photons. The rate of change in conductance is limited by the carrier lifetime, since the conductance, once changed, will remain changed until the excited electron recombines or is removed. Such devices with a dc bias have very limited bandwidth ($\sim$100 Hz). With a radio-frequency bias, the bandwidth can increase to several kilohertz. In view of the slow response, the photoconductive devices are not used for application in communications.

Photovoltaic Devices[7]

A photon absorption at a semiconductor *p-n* junction can give rise to the excitation of an electron into the conduction band, thereby forming a hole in the valence band if the photon energy exceeds the bandgap. This creates an open-circuit voltage, and a current will flow if the circuit is closed through a load resistor. If a reverse bias is applied to the *p-n* junction, the transit time can be made very small. The device will generate a current proportional linearly to the incident photon energy. A *p-n* junction diode of this type had varying characteristics at different illumination conditions, as shown in Fig. 6-2. At a constant bias the current is a linear function of incident optical energy.

In order to improve frequency response, the *p-n* junction is separated by an intrinsic region. This separation decreases the junction capacitance. Such a diode is often called a *positive-intrinsic-negative* (PIN) photodiode. An efficient PIN diode for high-frequency operation is made as small as practical to match the size of the optical beam spot that it is designed to detect, with its *p-n* junction close to the surface so as to minimize the amount of optical absorption before the *p-n* junction. The junction depth is made sufficiently large for photon absorption to be complete.

Under high-field conditions the reverse-biased junction can have secondary emission as the current flows across the junction. This can result in a noiseless gain just as that in the photomultiplier. This process is referred to as an *avalanche action,* and the diode is known as an *avalanche photodiode* (APD). For such a diode to be achieved, the ionization potential of the material must be relatively low while the

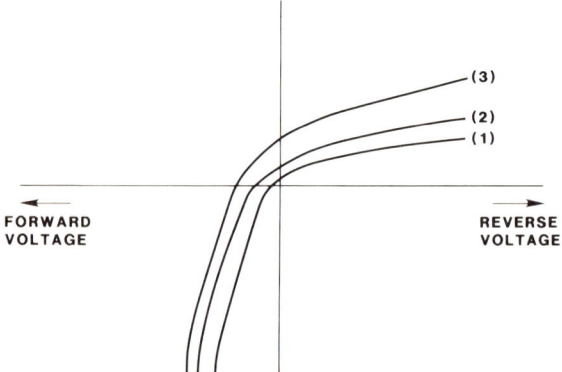

Fig. 6-2 Effect of different illumination conditions on a *p-n* junction diode: (1) no incident light; (2) and (3) increasing amounts of incident light.

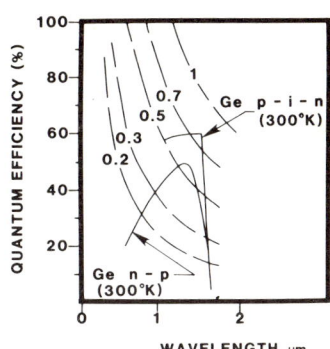

Fig. 6-3 Germanium response.

material uniformity must be high in order to avoid the possibility of local breakdown and plasma formation caused by local field intensity.

Material systems for photodiodes include germanium and silicon; InGaAs and InGaAsP are also possible detector materials. The spectral responses are as shown in Figs. 6-3 to 6-5. Frequency response can be derived from the signal equivalent circuit, as shown in Fig. 6-6. The cutoff frequency is given by

$$w = \frac{1 + R_s/R_p}{R_s C_p}$$

where

C_p = parallel capacitance
R_p = parallel resistance
R_s = series resistance
R_L = load resistance

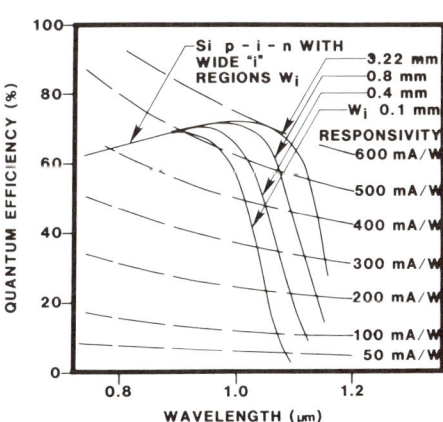

Fig. 6-4 Silicon response.

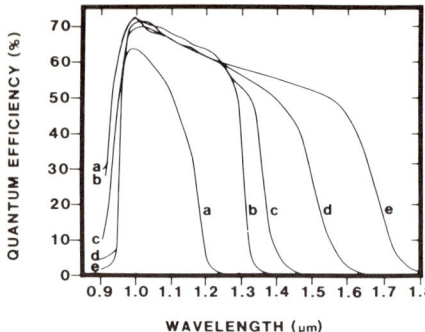

Fig. 6-5 InGaAsP responses a to e correspond to different quaternary compositions.

For diodes with area of the order of 10^{-2}mm, a cutoff frequency of the order of 1 GHz is readily obtained. The frequency response of an APD is more complex. In general, APDs can respond to gigahertz oscillations and maintain a gain in the order of as high as 100.

The high quantum efficiency, the broad bandwidth, and the fiber-compatible size make the semiconductor photodiodes ideal for optical fiber applications. The dark current for some devices is not negligible and must be taken into account in the receiver design.

The structures of a silicon PIN diode and APD are shown in Figs. 6-7 and 6-8, respectively.

Receiver Design Considerations[8,9]

Ideally, the receiver is to be designed to detect a minimum amount of signal power and reconstruct it to its original state with the addition of the least amount of noise and distortion. The fundamental factors giving rise to noise and distortion are to be examined.

A perfect receiver is one which generates electrons proportional to the incident optical power and that the electrons emitted can be tracked completely. Even in this situation, there is a fundamental noise component due to the statistical nature of the conversion process. The proba-

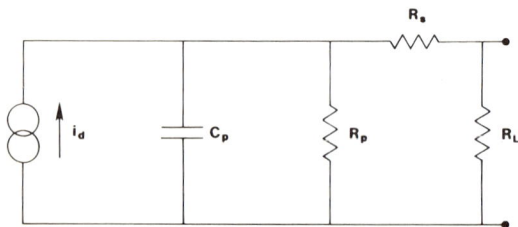

Fig. 6-6 Signal equivalent circuit.

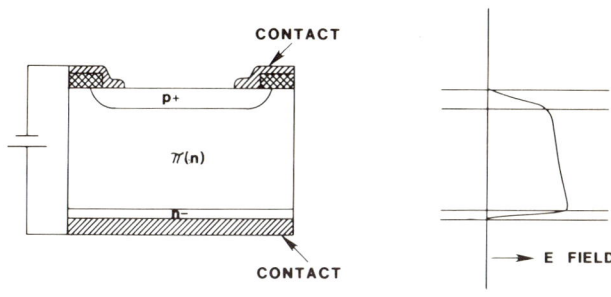

Fig. 6-7 PIN detector structure.

bility of k electrons emitted in response to a known incident photon energy in a time interval is given by the Poisson distribution:

$$P(k) = \left(\frac{\eta E_R}{h\nu}\right)^k \frac{\exp[-(\eta E_R/h\nu)]}{k!}$$

where $P(k)$ is the probability that k electrons are emitted in an interval, E_R is the total light energy incident upon the detector over this time interval, and η is the detector quantum efficiency, and $h\nu$ is the energy in a photon of frequency ν.

The probability for zero emitted electrons in the time interval is

$$P(0) = \exp\frac{-\eta E_R}{h\nu}$$

This is the probability of making an error in the idealized situation where the detection of one electron in the time interval concerned is considered as receiving a 1 and detection of no electrons is considered as receiving a 0 of a simple 1 and 0 binary digital signal. The finite

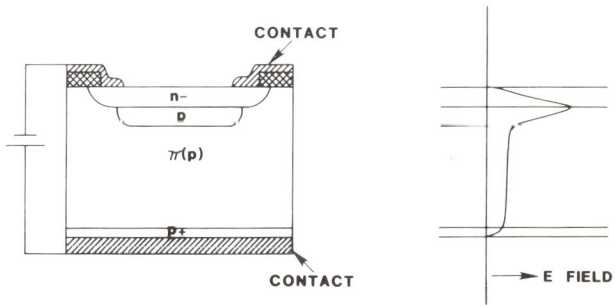

Fig. 6-8 APD detector structures.

probability of having no electrons emitted over the time with a given photon input is an error since the presence of a photon must constitute a 1. (Note: In this idealized situation the absence of photons constitutes a 0, and there can be no error since the detector cannot emit an electron with no photon input.) This is known as the *quantum limit*. For a given probability of error, a minimum amount of incident photon must be incident in a given time interval on the detector.

For an analog signal, it is possible to show that the SNR is $\eta P_R/(2h\nu\beta)$ in the presence of quantum noise only. The quantum-limit performance is approached when the detector contributes no noise.

In a practical receiver the electronics amplification stages following the detector contribute thermal noise. When the thermal noise is the dominant noise, the receiver design must minimize this noise so as to approach the performance limit set by the quantum limit. It is readily demonstrated in a simple circuit as shown in Fig. 6-9. This circuit is assumed to have a general-purpose amplifier of gain m.

If a single-electron charge from the detector appears at the resistor over a given time period, the induced voltage peak at the output can be shown to be about five orders of magnitude smaller than the thermal noise of the typical amplifier with 1 Ω at the input. In no way will the single electron charge be observable. It can also be shown that in the presence of the thermal noise, the reliable detectable number of photons in a time interval must be about five orders of magnitude larger than the number required for positive detection at the quantum limit condition.

It is possible to design specifically an amplifier to reduce the thermal noise effect. The noise output of a photodiode followed by the biasing resistors for the photodiode and an amplifier is lowest if the combined biasing resistors are very large. This results in a situation where the detector output appearing at the amplifier input is an integrated version of the signal. A matching equalization filter will be needed in order to compensate the integration by a differentiation process. This approach is known as the *high-impedance front-end amplifier design tech-*

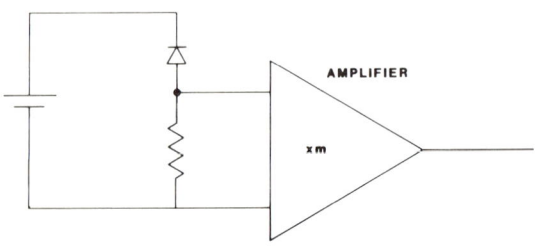

Fig. 6-9 Receiver model.

Modulation and Detection

nique and results in a low-noise receiver. The effective noise achieved by using a field-effect transistor (FET) amplifier which has a very high impedance input is about three orders of magnitude higher than that set by the quantum limit. For operating frequencies higher than 50 MHz, a bipolar transistor amplifier is required.

The situation can be improved by the use of the APD, since it can provide noiseless gain and thus can increase the number of available electrons per photon to combat the thermal noise. The APD, however, does introduce an additional noise term which increases with the gain, and for optimum low-noise design, this multiplication noise must balance the thermal-noise contribution. The analysis of minimum noise for an amplifier using an APD shows that for analog systems, improvement over the PIN detector case decreases as SNR increases. For high SNR, a quantum limit can be attained. This is shown in Fig. 6-10. For the digital case, optimization can also be reached; but because of the difficulties of predicting accurately the type of statistical distribution for the electrons, the calculation of error rate is complicated and less accurate.

In practical receivers the signal input is not always at one level. In fact, the amplifier must cater to an input variation of at least 1 to 10 or higher. In this situation the receiver design for lowest noise and highest sensitivity must be compromised to provide an adequate dynamic range. The high-impedance front-end amplifier has limited dynamic range in handling actual signal change as well as level changes due to signal pattern changes. A transimpedance approach has been used successfully to increase the dynamic range. This is a high-impedance amplifier with feedback. If the feedback resistor is large, the added noise is neglibible. Typically, a trade-off between poorer noise performance and larger dynamic range is possible. A 5-dB extra signal

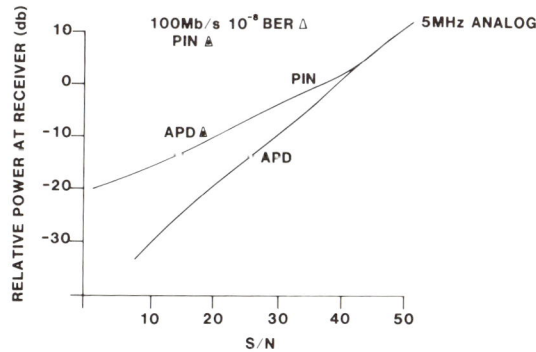

Fig. 6-10 Sensitivity comparison for PIN and APD receivers.

input is necessary to buy about 10 dB in dynamic range and to achieve an equivalent SNR or BER performance. A good amplifier for 45 Mb/s operation has a sensitivity of -55 dBm for 10^{-9} BER. The detected current at the input to the amplifier is in the order of 10^{-7} A. (Note: The drive current of the transmitter is around 10^{-1} A.) This means that the receiver must be well shielded from the transmitter in order to avoid internal crosstalk.

References

1. J. Brown and E. V. D. Glazier, *Telecommunications*, Chapman and Hall, London, 1969, Chapter 3.
2. Y. Yamamoto, "Receiver Performance Evaluation of Various Digital Optical Modulation-Demodulation Systems in the 0.5–10 μm Wavelength Region," *IEEE J. Quan Electron.* **QE-16**(11):1251–1259 (1980).
3. R. E. Epworth, "Phenomenon of Modal Noise in Fiber Systems," Optical Fiber Communication Conference, Washington D.C. (March 6–8, 1979), paper THD-1.
4. M. Schwartz, *Information Transmission Modulation and Noise*, McGraw-Hill, New York, 1970, Chapter 6.
5. H. Melchior, M. B. Fisher, and F. R. Arams, "Photodetectors for Optical Communications Systems," *Proc. IEEE* **58**:(10):1466–1486. (1970).
6. A. H. Sommer, *Photoemissive Materials*, Wiley, New York, 1968.
7. R. G. Smith, "Photodetectors of Fiber Transmission Systems," *Proc. IEEE* **68**(10):1247–1253 (1980).
8. S. D. Personick, "Receiver Design for Digital Fiber Optic Communication Systems," *Bell Syst. Tech. J.* **52**(6):843–886 (1973).
9. S. D. Personick, in *Optical Fiber Telecommunications*, S. E. Miller and A. G. Chynoweth, eds., Academic Press, New York, 1979, Chapter 19.

Chapter

Fiber Connectors, Splices, and Couplers

Introduction

Permanent splices and demountable connectors provide two means for connecting fibers. Couplers are required to achieve signal distribution. These devices may be realized with or without the aid of classical optical elements such as lenses and reflectors. The techniques to be employed must take into consideration the basic optical and physical characteristics of the fiber and the particular operational requirements.

In this chapter the relevant basic optical and physical characteristics of the fiber are summarized, and different types of connectors, splices, and couplers embodying various design principles are presented.

OPTICAL CHARACTERISTICS

When a fiber is cut, the fiber end is an open-ended waveguide. The near- and far-field patterns for single-mode and multimode fibers have all been studied in detail.[1] For connector design purposes, the radiation from the waveguide can be approximately described, with sufficient accuracy, using a modified geometric approach in which local acceptance angle or numerical aperture is defined for any unit area across the fiber aperture.[2]

The energy launched into a fiber can be computed according to the principles of field matching.[3] But for connector design purposes, the same geometric optics approach can be adopted to understand the launching conditions. Rays falling within the local acceptance angle are launched into the fiber, with the reflection loss given by the Fresnel reflection equation.

DIMENSIONAL TOLERANCES

The geometry of the core and the light distribution within the core of a fiber govern the dimensional precision required for the fiber to be joined to another fiber to achieve a low-loss connection. For example, if a 20-μm fiber core is to mate with another of similar dimension, a misalignment of 20 μm will cause the two cores to miss each other completely. If the light is confined within the core, a complete misalignment causes total loss of all the incoming energy.

In a circular cross-sectioned fiber waveguide, the relevant dimensions are (1) outside diameter (OD), (2) core diameter, (3) core ellipticity, (4) core concentricity, and (5) overall ellipticity. Typically, these can all be held to better than +1 percent if required. It is to be noted that the only dimension readily available for external reference is that of the OD of the fiber. Thus, for a fiber with 125 μm OD, with a tolerance of 1 percent, the maximum core-to-core misalignment, if the fiber is aligned off the OD, is 2×1.25 μm as a result of OD tolerance alone.

Additional control is required to align the axial directions and the end separation of the two fibers to be joined, as well as the quality of the end finishes. Assuming that two identical fibers are to be joined, the quality of the joint can be computed for different alignment tolerances.

JOINT LOSS

For identical fibers, a joint loss is the addition of the alignment losses and the reflection loss. The total loss can be derived by estimating geometrically, including reflection effects, the ray transfer from the input fiber to the output fiber. Three types of misalignment exist: end separation, axial displacement, and axial angular tilt.

In a practical case all three types of misalignment occur simultaneously. In addition, the fiber dimensional and optical tolerances cause additional misalignments. Thus, if the cores are aligned axially but do not match in absolute diameter, ellipticity, and numerical aperture, additional losses occur. However, the aggregate loss is a statistical addition of the effects of these misalignments. The maximum loss, corresponding to the addition of each loss component at its worst tolerance limits, is a rare case.

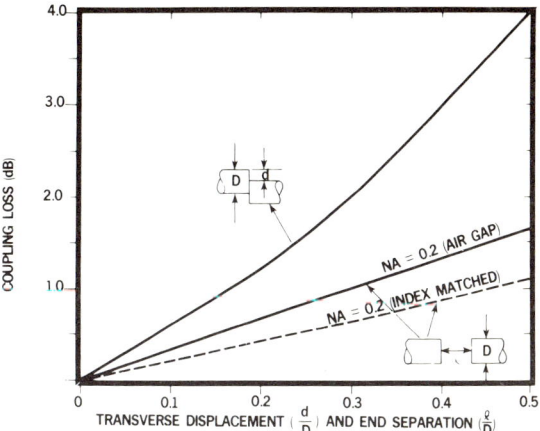

Fig. 7-1 Coupling losses of step-index fibers with transverse displacement and end separation.

EXAMPLES OF MISALIGNMENT LOSSES

Step Index

A fiber with a silica-germania core ($n = 1.464$) and a borosilica cladding ($n = 1.450$) has a numerical aperture of 0.2 and a maximum half angle θ_c of 11.7° in air ($n = 1$).

The computed misalignment losses[4] are illustrated in Fig. 7-1. It can be seen that the transverse misalignment is relatively more critical than end separation.

Loss due to angular misalignment is illustrated in Fig. 7-2, which shows that coupling efficiency is less sensitive to angular misalignment

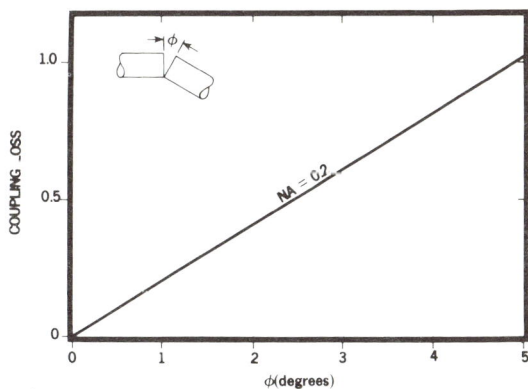

Fig. 7-2 Angular misalignment coupling loss.

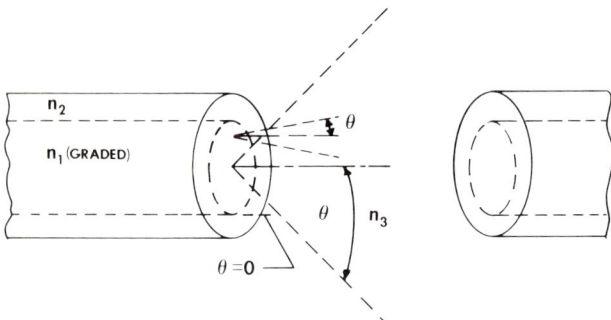

Fig. 7-3 Radiation from a graded-index fiber.

than the lateral alignment previously mentioned. Nevertheless, the angle between the axes of the fibers must be accurately controlled to within 1° for 0.2-dB loss.

Graded Index

The radiation pattern from the end of a graded-index fiber is different from that of a step-index fiber. The half angle of radiation from the transmitting fiber as a function of distance from the center of the core is illustrated in Fig. 7–3. As an example, a graded-index fiber that has a silica-germania core with a maximum refractive index (n_1) of 1.467 at the center of the core decreasing to a cladding index of 1.450 at the core-cladding interface has an effective numerical aperture of 0.22. At the center of the core, the maximum half angle θ is 12.9°. At the edge of the core the maximum half angle θ is 0°. Consequently, the coupling of a graded-index fiber is more sensitive to transverse misalignment than is a step-index fiber with the same core diameter. The coupling loss due to a lateral displacement of such a fiber is as shown in Fig. 7-4.[5]

End Preparation

For the connection of two fibers, the fiber ends must be smooth plane surfaces; otherwise, optical energy will be scattered or refracted, giving rise to modification of the angular distribution of optical energy and seriously increasing interconnection losses. The preparation of the fiber end is to achieve a plane smooth surface normal to the fiber axis.

There are two principal methods: the score-and-pull technique[6] and polishing.[7]

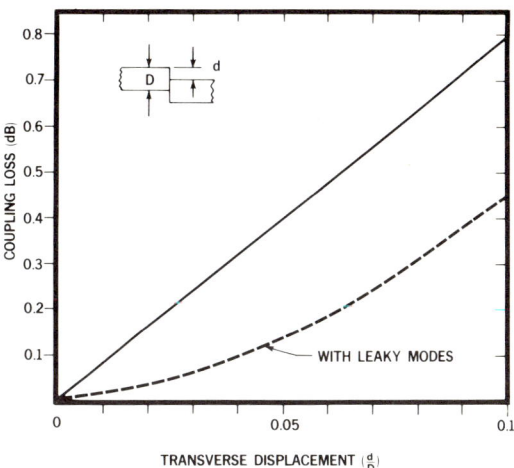

Fig. 7-4 Coupling loss of graded-index fibers with transverse displacement.

SCORE AND PULL

This technique involves scoring the fiber surface and then pulling to break the fiber. Fracture commences at the scored site. A mirrored finish can be achieved by the judicious control of the nature of the applied forces.

A fracture surface of a brittle solid is comprised of three regions known as the "mirror," the "mist," and "hackle" zones. The mirror zone is an optically smooth surface adjacent to the fracture origin, the hackle zone corresponds to an area where the fracture has forked and the specimen is separated into three or more pieces, and the mist zone is a transition region between these two zones. Figure 7-5 shows the fracture of a 125-μm glass fiber which clearly exhibits these three regions.

It has been demonstrated experimentally that the distance from the origin of fracture to a point on the boundary between the mirror and mist zones r is given by

$$Z(r)^{1/2} = K$$

where Z is the local stress at the point in question and K is a constant for a given material.

In order to break an optical fiber with the mirror zone extended across the entire fiber, it is necessary to maintain the stress at all points within the fiber so that $Z(r)^{1/2} < K$ is satisfied. The value of Z cannot be allowed to become zero or negative at any point across

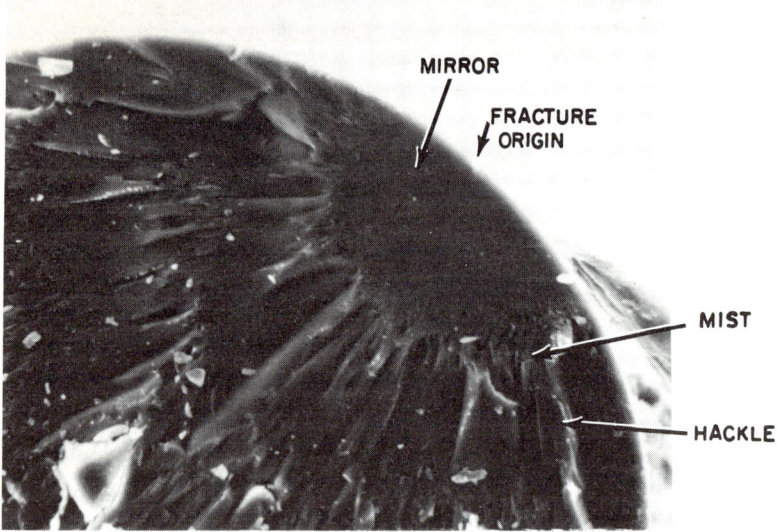

Fig. 7-5 Scanning electron microscope photograph showing origin of fracture of the fiber surface and the three types of fracture surfaces.

the fiber, or the crack will cease to propagate. It can be seen that to make a reliable clean mirror zone fracture, the stress distribution across the fiber must be adjusted. This can be done by bending the fiber into a radius R. The stress across the fiber, expressed as $Z(x)$ at $r = x$, is given by

$$Z(x) = T + \frac{E(a-x)}{R}$$

where T is the average tension on the fiber, E is Young's modulus, and a is the radius of the fiber (Fig. 7-6).

Fiber can be scored by using a sharp edge of a hard material such as diamond or alumina. Hot wire or electric arc can also be used effectively to induce an initial crack. In practice, the surface obtained by the score-and-break method is perpendicular to about 1° to the fiber axis.

POLISHING

Fiber ends may also be prepared by polishing. The fiber end to be polished should be embedded within a supporting material and held in a suitable jig to ensure that the finished surface is normal to the fiber axis. The supporting material should be sufficiently hard to prevent chipping of the fiber edge.

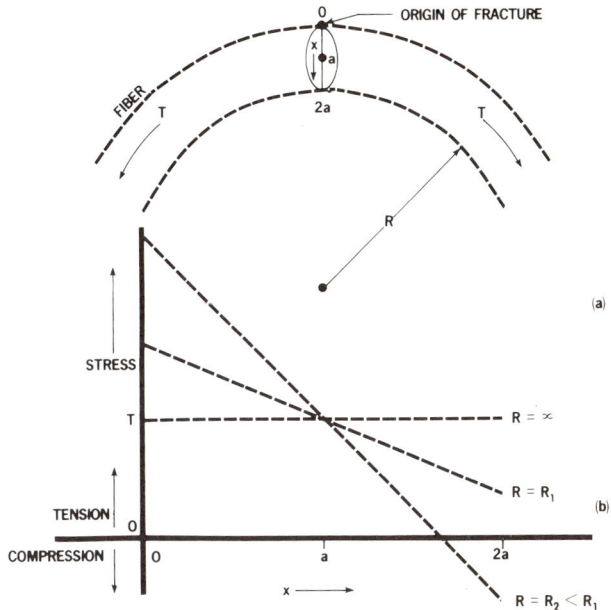

Fig. 7-6 Low-loss splices for optical fibers: *(a)* a glass fiber bent over a form of radius R and subjected to a tension T; *(b)* stress as a function of position in the fiber for various bending radii R.

Optical Fiber Splicing

An optical fiber splice is a permanent fiber-to-fiber joint. A splice is required in the fiber manufacturing stage as a means for increasing the fiber length and in the field as a way to make a joint or to repair a broken fiber. In the former case the spliced joint must retain the strength and dimension of the unspliced fiber. In the latter case the technique for making a splice should be simple and be possible to carry out in the field. An ideal fiber end for splicing is one which is flat and perpendicular to the fiber axis and that has a mirror-smooth finish.

FIBER FUSION SPLICE

A fiber splice suitable for silica fibers may be made by using an electric arc (Fig. 7-7), a plasma torch, or an oxyhydrogen torch to fuse or weld the fiber ends together.[8] Fibers with prepared ends are first aligned by micropositioners, and then heat is applied, fusing the ends together.

Mean losses of around 0.2 dB are readily achieved for multimode graded-index fibers with 50 μm core. Mechanical strength of fusion

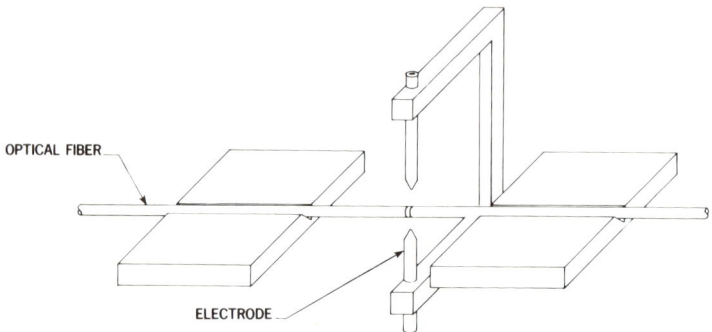

Fig. 7-7 Welding fibers with an electric arc.

splices can be made to withstand an average of 1 to 2 percent elongation. Fusion splice is readily implemented for both multimode and single-mode fibers. It is particularly suited to fibers with a lower melting and higher expansion core glass in a silica outer structure. In this case the surface tension forces help to align the fibers and also to ensure good contact between the two fiber ends. However, residue stress at the vicinity of the splice must be controlled. The exposed fiber can be recoated with plastic material or be enclosed within a protective tubular shield.

NONFUSED SPLICES

Techniques of splicing fibers without fusion can be envisaged to provide a convenient and rapid method of temporarily joining two fibers. If suitable bonding materials are used, these can also serve as permanent splices. The basic principles applied to this problem are to provide an alignment region of the required accuracy, easy means for allowing the fiber to enter this alignment region, and means for inserting a bonding agent to set the fibers in position. Some examples of these are given to illustrate the approaches.

Tube Splices[9]

A snug-fitting glass tube, a thermal-shrinking tube, and a glass sleeve with a center hole for insertion of adhesive are some structures which can be used as precision alignment tools. The ends of the tube are usually flared to facilitate the insertion of fibers. Epoxy or a thermoplastic is used to hold the splice together and also serves as an index-matching material.

GROOVED SUBSTRATES

Precision grooves of triangular cross section can be readily constructed on flat substrates. The dimension of the groove can be adjusted to accommodate the outside diameter (OD) of the fiber and allow alignment to be achieved. This technique allows easy splicing of a single fiber as well as an array of fibers.

One fiber optic cable splice utilizing this technique (Fig. 7-8) is formed by interleaving precision silicon chips with etched grooves and layers of fibers to form a two-dimensional array.[10] The array is then potted, ground and polished, to provide good ends on all the fibers. An array splice can be made by joining two such arrays in an alignment fixture. The average losses for splices assembled with these arrays have been reported to be in the range 0.2 to 0.3 dB, with a yield of 98 percent.

Optical Fiber Connectors

An optical fiber connector is a demountable fiber-to-fiber connection which serves to repeatedly make or break a connection. The principles adopted in connector designs fall into two basic categories: the precision butt joint approach and the optical wave-front transformer (lens) approach. The former uses practically achievable physical structures with the required lateral, azimuthal, and positional precisions to achieve fiber alignment. The latter employs lenses to assist in the alignment of the two fibers. It is aimed at trading the linear dimensional precision requirement for a more exact angular tolerance. Some examples of each type are to be discussed.

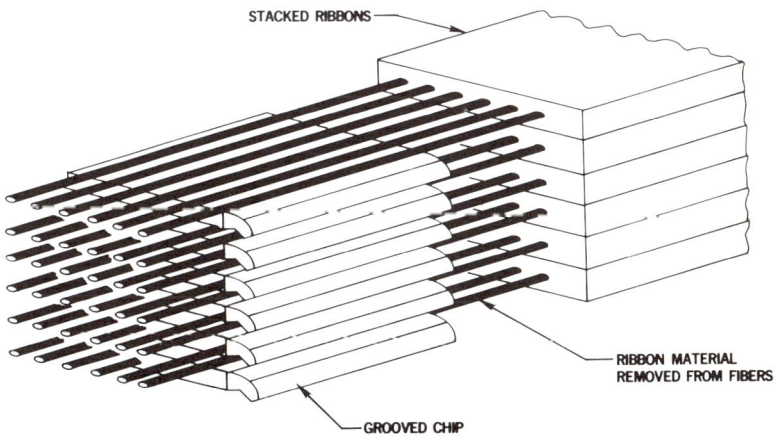

Fig. 7-8 Assembly of an array termination.

ECCENTRIC ADJUSTABLE CONNECTOR[11,12]

Accurate transverse alignment is achieved by rotating one fiber with respect to the other of two eccentrically mounted fibers until the monitored signal through the connector is maximized. Insertion losses of less than 0.4 dB have been achieved with single-mode fibers. The completed assembly can be locked into position permanently. The need to make initial adjustment is a distinct disadvantage. However, for a single-mode fiber, the precision achievable by this design is unique.

FERRULE CONNECTORS[13,14]

A rugged connector coupled with adequate, optical performance can be achieved by enclosing the fiber ends in protective precision ferrules (Fig. 7-9) and thereafter aligning the two ferrules. Care must be taken to ensure that the exposed fiber end is polished and slightly recessed from the end of the ferrule, thus greatly reducing the risk of damage during mating.

Maintenance of concentricity between the fiber and ferrule axes is of paramount importance, and the degree to which this can be achieved in practice largely determines the success of the finished connector.

PRECISION TRANSFER-MOLDED SINGLE FIBER CONNECTOR[15]

A process of transfer-molding a precision connector plug around an optical fiber can be used to achieve a fiber connector. Tapered surfaces are used to reduce the effect of abrasion and to facilitate insertion (Fig. 7-10).

A silica-filled epoxy is a suitable molding material which yields connectors of excellent mechanical integrity and abrasion resistance. The silica-filled epoxy has expansion coefficients sufficiently small to permit the connector to work over a wide temperature range.

The key part of the mold is a precision die which is cylindrically

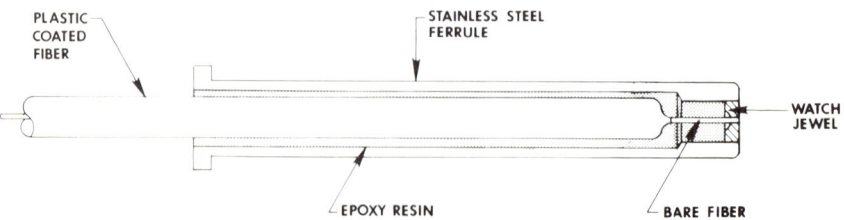

Fig. 7-9 Precision ferrule.

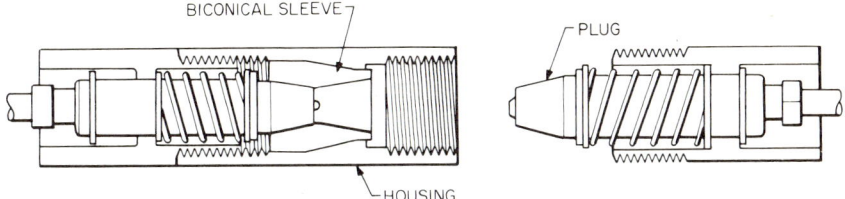

Fig. 7-10 Precision-molded single-fiber connector.

symmetric and has a centering guide hole which is concentric to better than a fraction of 1 μm with respect to the axis of the tapered surface. A biconical sleeve is used for aligning the connector halves. It is also formed by molding over a precision mandrel.

The achievable mechanical precision is indicated by the results of the transmission loss measurements. Figure 7-11a indicates extremely small variation in transmission loss with rotation of the sleeve. This indicated that less than 1 μm of eccentricity has been achieved.

Figure 7-11b shows that the variation of loss due to rotation of one plug corresponds to an achieved tolerance of about 1 μm. The distribution of average transmission loss, measured on 50 connectors, is plotted in Fig. 7-12.

Figure 7-13 shows the insensitivity of the connectors to temperature changes in the range of 0 to 60°C.

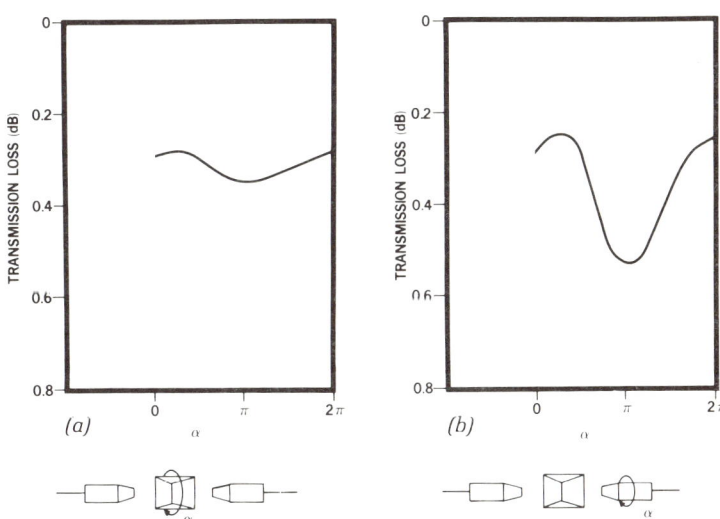

Fig. 7-11 Transmission variations of a molded connector.

128   Chapter Seven

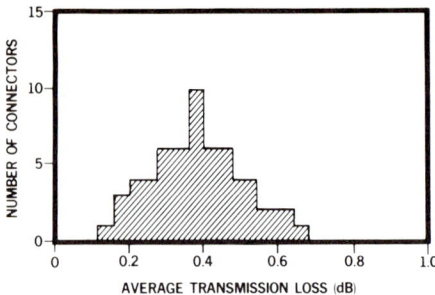

Fig. 7-12 Histogram of transmission losses.

LENS TRANSFORMERS

The use of lens transformers allows a trade-off of linear and angular tolerance requirements. In practice, the physical limitations of placing the fiber at the precise focal region of the lens, the size of the lens, the lens aberrations, and the increased number of reflecting surfaces are requirements which must be addressed in order to achieve a convenient and viable connector design.

A graded-index lens or an aspheric lens[16] can be used effectively. One successful design uses liquid as the lens[17] (see Fig. 7-14). A special molded form filled with the liquid of an appropriate refractive index provides the alignment mechanism for both the fiber and the lens structure. A repeatable performance of better than 1-dB loss has been reported.

An additional advantage of the lens-type connector is its ruggedness and resistance to dirt. Since the exposed surface within the connectors is much larger than the end of the fibers, dust particles appear smaller and can be cleaned off easily.

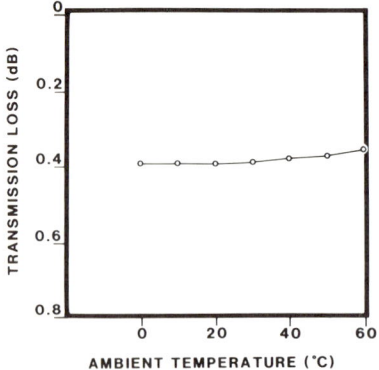

Fig. 7-13 Connector performance over a given temperature range.

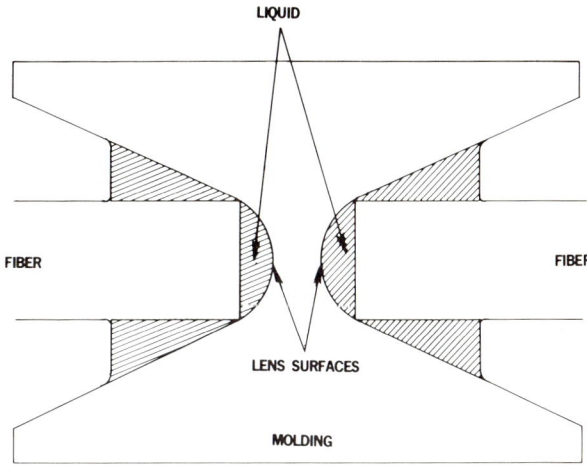

Fig. 7-14 Liquid lens transformer and fiber in position.

Couplers

Couplers can be defined as devices with three or more fibers interconnected to provide mutual coupling between them. In other words, a coupler is an N-port device with a defined scattering matrix. Functionally, couplers can be classified as directional, distributive, or wavelength-dependent couplers. The directional couplers may have either three or four ports. For power monitoring, the coupler may include a detector mounted at one of the ports of the coupler. Directional couplers may also be used as drop and insert elements along a data highway.

Distributive couplers are sometimes referred to as "star" couplers. In general, these couplers have an $M \times N$ structure, that is, M input ports and N output ports. The input and output ports may be separated, as in a transmissive distribution mixer coupler, or identical, as in a reflective distribution coupler.

It is useful to characterize passive couplers into three categories corresponding to the basic mechanism used: diffusion type, area-splitting type, and the beam-splitting type.

DIFFUSION COUPLERS

The coupling mechanisms employed in diffusion couplers are evanescent wave coupling and radiative coupling. In evanescent wave coupling, two or more fibers to be coupled are placed side by side in close proximity, such that the power of the guided wave in the input fiber is coupled gradually into the other fibers. This phenomenon is

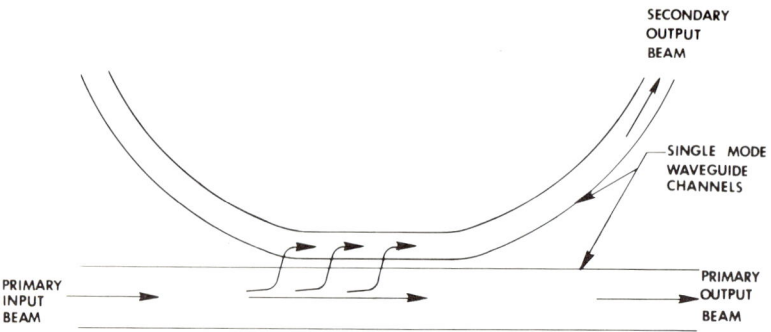

Fig. 7-15 Evanescent field coupler for single-mode fiber.

described in electromagnetic theory by the coupled-mode equations; an example of this is given in Fig. 7-15. In this case the coupling is through the evanescent field that extends deeply into the cladding region of single-mode fibers.[18] When two single-mode fibers are placed in parallel over a finite distance, the evanescent field from the primary fiber builds up a propagation field in the secondary fiber to provide two outputs. The evanescent field coupling cannot readily be used in multimode fibers, since the coupling is expected to be very weak and mode-selective.

In radiative coupling the power of the input fiber is made to radiate toward the output fiber. An example of this is given in Fig. 7-16. In this case the bent fibers couple to each other via the radiated field. Another important example of a coupler designed on the principle of radiative coupling is a simple and relatively efficient directional coupler for multimode fibers. This device is known as the *fused biconical taper coupler*[19] (see Fig. 7-17). The coupler is formed by twisting a pair of fibers which are then fused and elongated. Light entering the coupler along one fiber core is radiated out of the core due to the downtaper. This power is trapped in the cladding. Some of the trapped radiation

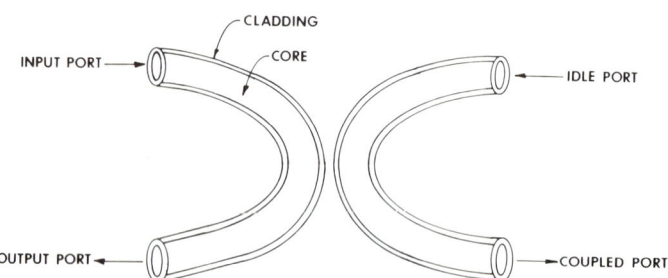

Fig. 7-16 Radiation-coupled bent-fiber directional coupler.

Fiber Connectors, Splices, and Couplers   131

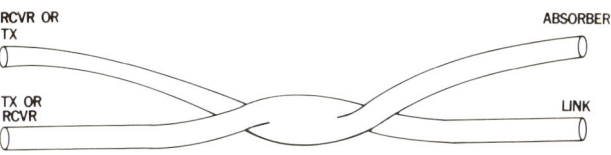

Fig. 7-17 Fused biconical taper coupler.

then reenters the core modes of each output fiber in the uptaper region. Excess losses, defined in terms of power radiated out of the coupler (core modes), can be as low as 0.5 dB.

AREA-SPLITTING COUPLERS

In area-splitting couplers the division of optical power within a guided structure is achieved by dividing up the light-guiding cross-sectional area. The schematic of a directional coupler shown in Fig. 7-18 shows the circular core cross sections characterized by the diameters D_1, D_2, and D_3 with

$$D_3 = D_1 + D_2$$

If the transition regions within the coupler, where the single-circle geometry ultimately transforms to two separate circles, are made to occur slowly and smoothly, the excess losses can be as low as 0.5 dB. The "star" couplers shown in Figs. 7-19 and 7-20 are two examples. This device distributes the optical signal from one of several transmitters to all the associated receivers as is required in a centrally distributed fiber optic data bus system. Figure 7-19 shows a transmission star,[20] and Fig. 7-20 shows a reflective star.[21] The transmission star separates the ports into transmission and receiving ports, whereas the reflective star does not designate ports into these categories. As a result, the

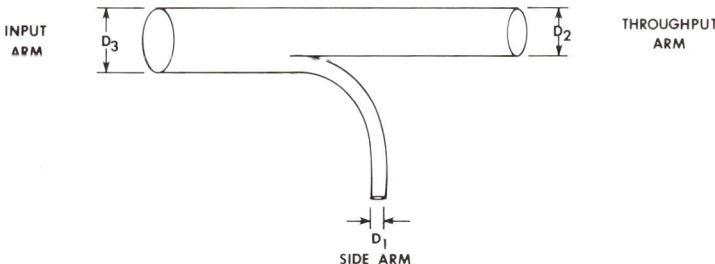

Fig. 7-18 Directional coupler.

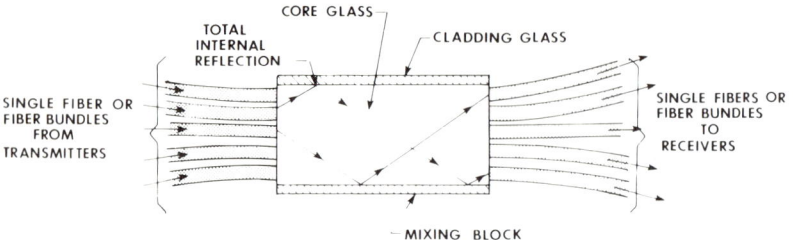

Fig. 7-19 Area ratio splitter transmission star coupler.

splitting losses are different, and the number of ports available for transmitters and receivers differs for the two cases.

Distribution stars can be formed by butting the fibers to a mixing rod. The fibers can be either plastic-clad silica or all-glass fibers. In the latter case the packaging loss is higher because the core-to-clad ratios are less than unity. If the interstices of the packing geometry can be eliminated by a carefully controlled deformation, much of the interstice packing fraction loss can be eliminated. A distribution star is required to have low excess coupling loss as well as good output uniformity from all ports.

In order to achieve output uniformity, the mixing region must have adequate length and should have a geometry which encourages mode mixing. Usually this is achieved by using a geometric shape other than a cylinder.

BEAM-SPLITTER COUPLERS

The third mechanism used in making couplers is partial reflection. The partially reflecting surface can be applied directly to the face of a fiber polished at an angle or can be in the form of a separate partially

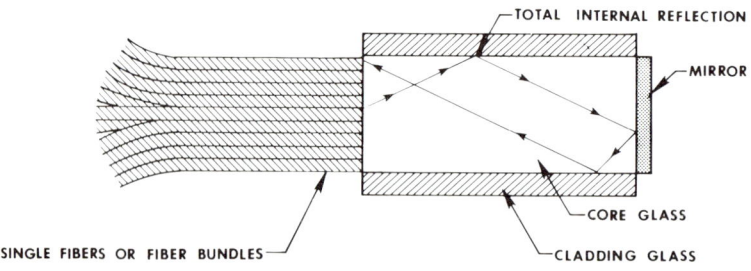

Fig. 7-20 Area ratio splitter reflecting star coupler.

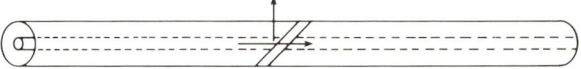

Fig. 7-21 Beam splitting by partial reflection.

reflecting mirror. One monitor coupler[22] for optical sources is as shown in Figure 7-21.

In Fig. 7-22 the beam splitter is achieved by applying a semitransparent coating to a pair of graded refractive-index lenses[23] (Selfoc lenses), in place of the dichroic reflector.

WAVELENGTH-SELECTIVE COUPLERS

If a dichroic reflector is used in place of the partially reflecting mirror, a wavelength-selective coupler is achieved.[23] The dichroic surface is used for wavelength selection, so that the coupler enables efficient full duplex operation on a single-fiber channel to be accomplished. This type of coupler can also be used for two or more wavelengths to be directed into a single fiber and thereby achieve multiplexing through the use of two or more wavelengths.

Figure 7-23 shows the basic configuration of three wavelength-selective mechanisms: dispersive prism, dispersive grating, and multilayer dielectric interference reflector with dichroic properties.

An analysis of the capabilities of these wavelength-selective mechanisms leads to the following conclusions. The prism coupler has a slight advantage in throughput loss compared to the dichroic, while the latter is potentially somewhat better in backscatter characteristics. Both the prism and the dichroic appear to be superior to the grating coupler. The advantage of the grating coupler over the others is its higher wavelength-separation power.

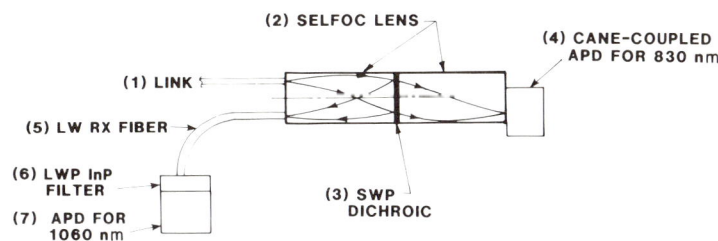

Fig. 7-22 Bulk dichroic beam-splitter bidirectional wavelength duplexing coupler.

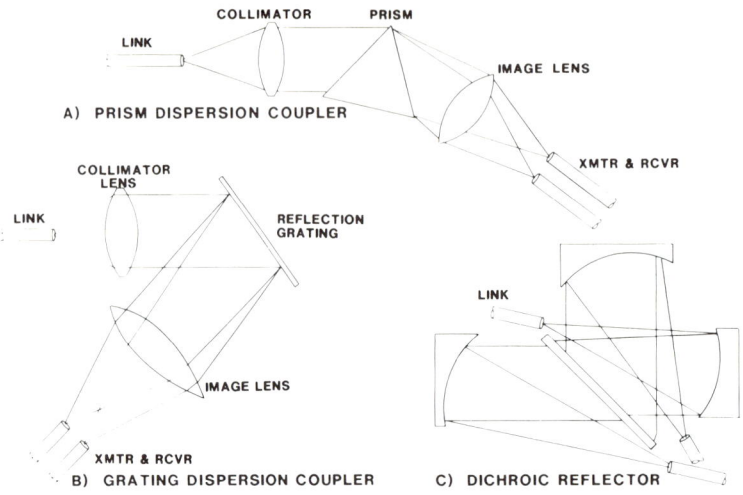

Fig. 7-23 Wavelength selection mechanisms.

References

1. M. Imai, "Average Intensity Distribution of Far-Field Radiation Patterns in a Multimode Optical Fiber," *Transact. Inst. Electron. Commun. Eng. Jap., Sect. E* **E63**(1):16–23 (1980) (14 references cited).
2. D. Gloge, "Propagation Effects in Optical Fibers," *Transact. IEEE* **MTT-23**:106–120 (1975).
3. A. W. Snyder, "Excitation and Scattering of Modes on a Dielectric or Optical Fiber," *Transact. IEEE* **MTT-17**(12):1138–1144 (1969).
4. K. Miyazaki, M. Honda, T. Kudo, and Y. Kawamura, "Theoretical and Experimental Consideration of Optical Fiber Connector," Topical Meeting on Optical Fiber Transmission, Williamsburg, Va. (January 1975), paper WA4–1.
5. T. C. Chu and A. R. McCormick, "Measurements of Loss Due to Offset, End Separation and Angular Misalignment on Graded Index Fibers Excited by an Incoherent Source," *Bell Syst. Tech. J.* **57**(3):595–602 (1978).
6. D. Gloge, P. Smith, D. Bisbee, and E. Chinnock, "Optical Fiber End Preparation for Low-Loss Splices," *Bell Syst. Tech. J.* **52**(9):1579–1588 (1973).
7. C. Wilson and P. J. Stevens, "Polishing Technique for Optical Waveguide Terminations," *J. Phys. E. (UK)* **7**(8):614–615 (1974).
8. I. Hatakeyama and H. Tsuchiya, "Fusion Splices for Optical Fibers by Discharge Heating," *Appl. Opt.* **17**(12):1959–1964 (1978).
9. C. Miller, "Loose Tube Splices for Optical Fibers," *Bell Syst. Tech. J.* **54**(7):1215 (1975).
10. C. M. Milleo, "Fiber Optic Array Splicing With Etched Silicon Chips," *Bell Syst. Tech. J.* **57**(1):75–90 (1978).
11. M. Borner, D. Gruchmann, J. Guttman, and O. Krumpholz, "Detachable Connector for Monomode Glass-Fiber Lightwave Guides," *Archiv. fur Electronische and Ubert.* **26**(6):288–289 (1972).
12. H. Tsuchiya, H. Nakagome, N. Shimizu, and S. Ohara, "Double Eccentric Connectors for Optical Fibers," *Appl. Opt.* **16**(5):1323–1331 (1977).
13. J. D. Archer, "Single Fiber Optical Connections," *New Electron.* **11**(2):51 (1976).

14. N. Suzuki, Y. Iwahara, M. Saruwatari, and K. Nawata, "Ceramic Capillary Connector for 1.3-μm Single-Mode Fibers," *Electron. Lett.* **15**(25), 1979.
15. P. Runge, L. Curtis, and W. Young, "Precision Transfer Molded Single Fiber Optic Connector and Encapsulated Connectorized Devices," Topical Meeting on Optical Fiber Transmission, Williamsburg, Va. (1977), paper WA4-3.
16. J. S. Leach, M. A. Matthews, and D. Dalgoutte, "Optical Fibre Cable Connections," in *Optical Fibre Communication Systems*, C. P. Sandbank, ed., Wiley, New York, 1980, Chapter 15.
17. M. Holzman, "Detachable Connectors for Multi-Mode Graded Index Optical Waveguides," Digest of Technical Papers, Conference on Laser and Electro-Optical Systems, San Diego, Ca. (February 1978), paper WAA4.
18. S. K. Sheem and T. G. Giallorenzi, "Single-Mode Fiber Multiterminal Star Directional Coupler," *Appl. Phys. Lett.* **35**(2), 1979.
19. B. S. Kawasaki and K. O. Hill, "Low Loss Access Coupler for Multimode Optical Fiber Distribution Network," *Appl. Opt.* **16**(7):1794 (1977).
20. E. Rawson and A. NaFarrate, "Transmission Star Coupler for Single Fiber Cables; Mixer Rod Power Distribution Inhomogenieties," Digest of Technical Papers, Conference on Laser and Electro-Optical Systems, San Diego, Ca. (February 1978), paper WFF.
21. D. C. Johnson, B. S. Kawasaki, and K. O. Hill, "Low-Loss Reflection Star Couplers for Optical Fiber Distribution Systems," *Appl. Phys. Lett.* **35**(7), 1979.
22. H. Kuwahara, J. Hamasaki, and S. Saito, "A Semi-Transparent Mirror-Type Directional Coupler for Optical Fiber Applications," *Transact. IEEE-MTT* **21**(1):179–180 (1975).
23. K. Kobayashi, R. Ishikawa, K. Minemura, and S. Sugimoto, "Micro-Optics Devices for Fiber Optic Communications," *Fiber Integrated Opt.* **2**(1):1–17 (1979).

Chapter

Systems

Optical fiber systems are divided into three types: transmission, distribution, and sensor. The characteristics of the fiber can be exploited to achieve systems which are economically competitive against other solutions while offering simultaneously extra advantages and unique functions at an affordable cost.

Transmission systems are point-to-point links. Optical fiber with its low loss and high bandwidth enables the realization of systems of large information-carrying capacity and long span length between repeaters. Optical fiber transmission systems operate with advantage for many different applications. The more important uses are:

1. For the public telephone/data trunk network:
   - interoffice link (interconnecting central offices)
   - intercity link (interconnecting central offices between cities)
   - entrance link (satellite to hub)
   - central office to remote switch center link
   - 24 or 30 channel pulse-code modulation (PCM) links for high-electromagnetic-radiation field environment
   - undersea and long-haul system
2. For the cable TV network:
   - cable TV trunking
3. For improved privacy:
   - tamperproof link
4. For military applications:
   - interbase link
   - entrance link
   - mobile link

- radar remote
- weapon guidance
- intravehicular transmission

Distribution systems are multiuser-multiservice systems aimed at singly and multiply interconnected mode of usages. Optical fiber, with its broad bandwidth, can realize such systems with increased flexibility and capability. The more important systems are:

1. For the integrated multiservice network:
   - wired office
   - wired city
   - computer network
2. For the cable TV network:
   - cable TV distribution

Sensor systems employ a fiber not as a transmission medium, but as a sensor. The ability of single-mode fibers to transmit coherent radiation and retain the phase information enables fibers to be used in sensor systems, for example, as a gyroscope, an acoustic sensor, or a temperature sensor. In this chapter a number of selected systems from each of these three types of systems are discussed. For each, some aspects of the design approach, the economic considerations, and the performance advantages are highlighted.

Interoffice Link[1]

The term "interoffice link" is a generic name for a system connecting two telephone central switching offices. The traffic between the switching centers ranges from tens of channels to many thousands of channels. The distances involved range from a few kilometers to more than 10 km. In a telephone network many such links are required. Transmission is usually arranged to handle a group of channels together. In the United States the groups are called T1-24 channels, $T2 = 4 \times T1 = 96$ channels, and $T3 = 7 \times T2 = 672$ channels, while the European standards are at 30 chan, 120 chan, and 480 chan, and the Japanese are 24 chan, 96 chan, and 480 chan. Even higher groups have been designated.

A low-loss, large-bandwidth optical fiber offers a great deal of flexibility in designing optical fiber systems to carry the chosen group of channels between the central offices. Obviously, the system installation and maintenance can be greatly simplified if no repeaters are used between the offices. The equipment can also be placed in a more benign environment provided by the central office instead of the harsher climate in an external repeater housing. The designed capacity is dependent on the actual system requirement. The cost-effectiveness of the link, however, is dependent on the capacity. For low-capacity operation, the transmitter, receiver, and cable costs must be kept low in order to compete with copper systems. For high-capacity operation, optical systems are economically very attractive compared with other types of systems.

The performance advantages of the optical fiber, such as small size, light weight, low loss, large bandwidth, are translated into providing ease of installation, long repeaterless span length, and adequate traffic-handling capacity. These features, in turn, can be translated into cost savings in specific cases. For instance, suppose that a metropolitan interoffice link is to be installed in a congested duct. The small-sized fiber cable may be pulled into the available space with cable-pulling equipment which can be maneuvered in confined spaces. The absence of repeaters means that no special housing is required along the route and thus could represent very significant cost savings. It is certainly understandable why this type of system was the first to be implemented in large scale.

Entrance Links

The satellite ground stations and radio-link terminals (headend) are located in relatively uncongested areas away from city centers. These are stations where large volumes of traffic converge. The interconnection distance of these stations to central offices is of the order of 10 km. Optical fiber can provide a convenient link and can relieve the radio-frequency spectrum congestion which would arise if the connections to the city center are via radio links. Optical fiber entrance links ideally should provide repeaterless connections, although the use of a repeater is not out of the question.

The signals are usually in frequency modulation (FM) format. They are on a microwave carrier and are demodulated to a 70-MHz intermediate frequency (IF) carrier. The actual signal may be a TV signal or a digital bit stream of telephone and data messages. A possible way of handling the analog signal is to use the signal on the IF carrier to intensity-modulate an optical carrier. The IF band extends to over 90 MHz, and because of the span length to be achieved, a laser transmitter is usually required. For a digital signal, it is often more easily handled after demodulation from the IF to the baseband.

By transmitting the signal on the IF carrier, the economics of the entrance link is attractive, but due to the high SNR required, the span length is limited to a few kilometers. By demodulating the signal and transmitting the digital signal at baseband, the terminal equipment may be more expansive, but the transmitter linearity and the receiver bandwidth are less demanding, and longer span length is easily achieved. The equipment required is then similar to the interoffice link. The system economics can also be very favorable.

24 or 30 Channel PCM Link for a High-Electromagnetic-Radiation Field Environment[2]

This is a specialized product designed for use in areas where high electromagnetic radiation can cause interference to the transmission system. In a high-voltage transformer site, the induced current on a copper telecommunications cable can be sufficiently high to melt the cable. In such an environment even a shielded cable is not usable, since the shielding will act as the current conductor. An all-dielectric optical fiber can be placed in such an area to establish the communications link. Along an electrified railway the thyristor-controlled motors generate interference signals of very large magnitude. Doubly shielded copper cable may prevent the interference from reaching the signal-carrying conductors, but there is no 100 percent assurance that the shielding will be effective at all times. Optical fiber cable can again be used to establish the vital communications link associated with the efficient operation of railways. For high-voltage grids, it is often useful if information transmission lines can be carried along the power lines to carry the signals associated with the monitoring of the power systems. Optical fiber cables can do just that.

The 24 or 30 channel PCM link is designed to transmit a 1.5- or 2.0-Mb/s bit stream. At this bit rate a copper wire system is often more economical than the optical fiber version. However, for use in the high-electromagnetic-radiation-field environment, the fiber system can prove out against a copper wire system employing shielded cable design and may be the only type of system capable of working in that environment.

In most applications of this type the link span length is relatively short. This means that an LED light source and a PIN detector can provide the required system margin. If long span length is required, as along the high-voltage power grids, a laser source and an APD receiver can give a span length of >10 km at 0.85 μm and >50 km at a longer wavelength.

Undersea Long-Haul Systems[3]

Undersea cables formed the only long-distance communications link across the oceans in the days before the satellite system was introduced. Since then, undersea cables have been shown to continue to play a dominant and important role with satellite systems, serving as an important complementary facility for worldwide communications needs. Furthermore, undersea systems designed to link up with offshore equipment and to function in many geophysical explorations across a stretch of water have found increasing use.

Optical fiber, with its favorable size-to-strength ratio, adequate bandwidth capability, and low loss, has been recognized as a good candidate for application in the undersea systems. The optical fiber undersea cables can be designed to survive such an operational environment, with a design which has smaller overall cross-sectional size and a repeater span length longer than what a copper cable can achieve for the same information-carrying capacity. These factors provide economic savings in cable cost and cable laying cost. The fewer repeaters allow the more complex electronics for handling digital information to be introduced, without sacrifice to system reliability. For special applications, such as geothermal experiments and tow cables for instrumentation packages at sea, optical fiber can readily be incorporated into these special cable structures and can meet the high stress and/or high temperature requirements besides providing improved transmission characteristics.

DESIGN CONSIDERATIONS

For long-haul trunking operation, undersea long-haul systems are designed to carry substantial information traffic over distances up to transoceanic lengths. The design is aimed at achieving the largest information-carrying capacity and over the longest span possible. This means that large-bandwidth single-mode fiber is the preferred type of fiber. Operation wavelength is chosen to coincide with the lowest loss region of the fiber spectral loss. Two windows are attractive, one at about 1.3 μm and the other at about 1.6 μm. The 1.3-μm region has a slightly higher loss but has almost zero material dispersion, which can be compensated by the waveguide dispersion easily. When the 3 to 5-μm region is eventually ready with a suitable fiber, source, and detector, the potentially achievable 0.01 dB/km fiber loss allows a transoceanic repeaterless system to be envisaged. It is to be noted that at a system length of 5000 km, the maximum bandwidth is limited by either the material or the waveguide dispersion. For a larger bandwidth to be obtained, the waveguide dispersion must be designed to cancel the material dispersion.

The transmitter employs a single transverse mode and preferably a single longitudinal mode laser with a small emission region to match the single-mode fiber. The choice of a low-threshold-current, small-sized source with narrow spectral emission linewidth facilitates in the design of a transmitter with maximum coupled power to the fiber and with minimum required drive power. Minimization of power consumption is of paramount importance in undersea cable systems incorporating repeaters, since it can simplify the power supply requirements. A low power consumption at the repeater reduces the copper wire cross-sectional dimension for power carrying purposes and allows local powering via some form of storage batteries for special applications.

The receiver is designed for maximum sensitivity but is compatible with low power consumption and minimum circuit complexity. Nonregenerative amplifiers should be used interspersed with regenerative repeaters when timing jitter and pulse distortion become sufficiently large and are in need of correction.

With a 0-dBm transmitter power available at the input to the fiber and a receiver sensitivity of -55 dBm, the maximum span length—assuming a fiber loss of 0.5 dB/km and splice loss of 0.1 dB at 10-km spacing and no connectors—for a 100-Mb/s signal is 100 km. The single-mode fiber dispersion must be 10 Gb/s·km.

The fiber for the undersea cable must have good dimensional stability for low splicing loss and low mode-conversion loss. The fiber numerical aperture is likely to be <0.2 in order to achieve the low loss. This means that due attention must be paid to the fiber dimensional stability and packaging, especially if the full water pressure is allowed to be applied to the fiber. Designs to shield the fiber from water pressure, however, are available. For shallow-water application, the provision of adequate armoring is also necessary.

Perhaps the most serious concern of an undersea optical fiber system is the reliability of the electronics. This implies the development of highly reliable optical sources, detectors, and the semiconductor digital circuits which are available in both discrete and large-scale integration (LSI) forms. The temperature range expected is relatively small. This is a welcome relief. As far as the fiber cable is concerned, the fiber durability can be made adequate by ensuring low residual stress. However, care must be taken in the choice of material and application method of the protective coating. An extra criterion the fiber is required to meet is the hosing effect in the case of severance of the cable. Under high water pressure many of the coatings will allow water to flow between the interface, thus resulting in the coating acting as a water hose.

For single-hop applications, a span length of 100 km can be used advantageously to interconnect islands and to cross straits and channels. In such applications the reliability of the electronics required is the same as that commonly encountered with land-based systems and thus poses no problems.

Cable TV Trunking

Cable TV is the distribution of TV signals to subscriber homes by cable instead of by free-space radiation. The central distribution points are called *hubs,* which receive TV signals from TV stations directly or via the headend at which TV signals are received from free space and from satellites. The transportation of the signals between headend and hub and between hub and hub is referred to as *trunking*. Trunking is carried by special low-loss coaxial cables with the TV signal assembled in a multiplexed form. Some systems carry 12 channels; most systems carry at least 20 channels, and others carry 40 or more channels. The repeater spacing is about 1 km, and the number of repeaters is limited by noise and distortion to around 10 unless special equipment—for example, equipment employing frequency modulation—is used. For trunking over long distances, an optical fiber system is a possible alternative solution.

The economics involved in a trunk route depends on the capital cost, the installation cost, and the maintenance cost. The coaxial system is very cost-effective, since the number of channels the amplifiers can handle is large, and the coaxial cable also has adequate bandwidth. The installation cost is reasonable, since repeaters are pole-mounted and have no additional enclosures. However, the maintenance cost is high. The temperature variations cause gain and equalization variations, often in excess of the range of the automatic gain and equalization circuits, resulting in the need for maintenance crews to perform manual adjustments. Furthermore, amplifier reliability is required to be high; otherwise, service availability could be too low to be acceptable.

Optical fiber can be used for cable TV trunking purposes, provided that separate fibers are used to carry the channels in digital form. The advantage is to obviate the need for a repeater, thus yielding improved reliability and low maintenance cost. Another advantage is the ability to cope with longer trunking distances. In fact, with long-wavelength operation, a trunking distance of 30 km can be achieved without a repeater. The ecomomic disadvantage, of course, is the need for trunking of as many fibers as there are TV channels. This situation can be relieved somewhat by suitably processing the TV signal and/or the use of wavelength multiplexing. The TV signal bandwidth can be compressed by a factor of at least 2, thus enabling two channels to be handled by one fiber instead of two. This solution requires a more complex TV signal digitizer but calls for just one transmitter and receiver. The use of wavelength multiplexing increases the number of channels a fiber can handle, but the number of transmitters and receivers must also increase. Both schemes can be applied simultaneously. The trade-off is between the electronics processing cost versus the fiber cost.

DESIGN CONSIDERATIONS

The signal-to-noise (SNR) requirement of an analog TV signal is marginally met on an optical fiber system over a system length of 1 to 2 km. It is necessary to convert the analog signal to digital format before satisfactory trunking over longer distances. This conversion can be implemented without degradation of the signal-to-noise ratio (SNR). The transmitter-receiver design for digital TV is similar to the case of the interoffice link, but with the digital bit rate raised to around 100 Mb/s. This does not present much additional technical difficulty in comparison to the 45-Mb/s case.

For trunking applications, the repeaterless solution is most attractive. For nationwide trunking, the optical fiber solution is most appropriate but must compete with trunking via the satellite. For smaller-scale trunking applications, the improved reliability and reduced maintenance cost can be important, especially when the cost of digitization of TV signal with compression decreases.

Tamperproof Link

The privacy of a telecommunications system depends on whether the signal along the transmission line can be remotely detected outside the line. Copper cable systems operate at frequencies where electromagnetic coupling can be effected by placing field-coupled line at a point relatively remote from the cable. This means that the privacy of a copper system is not high. At optical wavelength the electromagnetic field coupling is possible within distances equal to several wavelengths. In practice, this means extremely close proximity; thus optical fiber systems have improved privacy.

It is possible to induce radiation from a fiber, for example, by introducing a bend of small radius ($\leq$ 1 to 2 mm), but this implies physical access to the fiber. Steps can be taken to prevent physical access, and schemes can also be envisaged that permit the detection of physical tampering with the fiber.

DESIGN CONSIDERATIONS

A system with a high degree of privacy should be designed so that the signal level within the fiber is kept at a level just adequate for low error-rate detection at the receiver. This allows tampering to be detected simply because it will cause a signal-level decrease in the system, resulting in increased error rates at the receiver.

Military Applications[4]

The communications needs of the military parallels those of the nonmilitary public sector. The communications requirements within a military base resemble those within an office complex, while interbase systems are similar to interoffice links. There are also entrance links from head-end to hub. Optical fiber applications in these areas provide the advantages. Apart from signal format details, the designs are generally similar.

There are other military applications, of mainly tactical types, which explore more fully the special feature of optical fiber, such as small size, light weight, flexibility, and strength, to achieve systems fulfilling military needs. Some of these are to be discussed.

Mobile Link

It can easily be envisaged that a communications network is vital between locations serving as temporary bases for the military. Such locations are unlikely to have communications installations, and even if there were installations, these might not be functional. A quick means of establishing such a communications link is important. The efficiency of carrying out such an installation is dependent largely on the size and weight of the equipment and the method of deployment. The cable for the system must be carried around and thus would present a logistics problem and would prevent rapid installation if its size and weight were large. The use of a lightweight but strong optical fiber cable can be a major advantage. In fact, fiber cables can be laid from small vehicles or even from backpacks or helicopters. The electronics parts and additional lengths of cable must be readily interconnectable, preferably via dirtproof connectors.

DESIGN CONSIDERATIONS

Designing for portability and ease of installation, the transmitter-receiver should be battery-powered and have low power consumption. To provide easy connection to the fiber cable, a single connector for a duplex link is needed. This same connector should be usable for cable-to-cable connection as well. The cable should have minimum weight and size compatible with the required ruggedness. An all-dielectric structure is preferred since it will not be a radar target.

Cable Design

An all-dielectric cable that uses Kevlar yarn or glass fibers as the strength member has a high strength-to-weight ratio. This enables the weight and cable cross section to be minimized. The outer sheath must be highly resistant to abrasion and be inert to most chemicals. A material such as polyurethane has properties close to these needs, and can be successfully used as the sheath material. The choice of sheath material imposes a certain amount of constraints on the choice of fiber-coating materials and/or cable design, since the sheath extrusion must not melt the material within the sheath.

A practical fiber cable can be designed to withstand the pressure and shear forces created by a tank driving over it and to have a diameter sufficiently small and a weight sufficiently low to allow for a kilometer length to be carried on a backpack for manual deployment.

Connector Design

The operational environment of a military product is hostile. The connector is required to provide low-loss connections, even after it is dropped into muddy water, is blown in sandblasts, and has suffered other forms of extreme treatment. If a butt-joint-type connection is envisaged, the extremely tight lateral dimensional tolerance must be maintained and surface cleanliness preserved during operation. This calls for designs which allow the displacement of dirt while the connector halves are being mated and fiber ends protected when the connector halves are separated. These features are in addition to the need to achieve very high lateral tolerance (in the order of <5 percent) of the fiber core diameter. For operational ease the connector is to be hermaphroditic, that is, capable of being mated without having to have a plug and receptacle of different noninterchangeable design. The size of the connector should be sufficient for easy handling by personnel in arctic conditions when large, inflexible gloves prevent fine manipulation.

In one possible design the lateral tolerance requirement is facilitated by the use of a specially designed lens. The lens spreads the light from the fiber such that a small misalignment does not cause undue loss between the mating connector pairs. The use of a lens also imposes a less stringent demand on the surface perfection. In fact, this connector should be able to withstand the operational conditions without having to cover the lenses when the connector is separated. If dropped in dirt, a simple wipe clean may be acceptable. The allowable separation between the lenses also prevents the lens surface from being abraded by the trapped dirt particles.

Deployment

Apart from laying the cable on foot, the use of motorized vehicles is naturally even more convenient. A comparison with the equivalent copper cable system shows that fiber systems provide great savings in logistic support needs. Finally, helicopter laying of a lightweight cable is a distinct possibility. This can be performed with much higher speed on open grounds and possibly over water as well.

Radar Remote

A radar is an active reconnaissance device consisting of an antenna and its signal processing equipment. Because the antenna site radiates electromagnetic energy, it is easily detected and may come under attack. The signal analysis equipment is complex and sophisticated and often is expensive and bulky. During hostilities, the radar antenna should be placed as far removed as possible from the signal analysis equipment, which usually is a manned post. In this way the personnel involved are exposed to less danger, and expensive equipment can also be hidden more strategically. Furthermore, the antenna unit can be made more mobile and be deployed more easily.

DESIGN CONSIDERATIONS

For sophisticated radars such as the phased-array antennas, the signal characteristics must be preserved not only in amplitude, but also in the phase. This means that the signal is analog in nature and must be handled accordingly. The transmission characteristics of the optical fiber link must be linear in order to avoid distortion of the information and should have a very large bandwidth in order to achieve highest time, and hence target, resolution.

Weapon Guidance

The accuracy of long-range artillery firing has been partially improved by computerized estimation of the trajectory, taking into account range, wind velocity, and other relevant factors. This form of weapon control is routinely practiced in order to improve the statistical probability of making the weapon more effective. In the case of major weapons, such as missiles and torpedoes, the incentive to improve accuracy is greater: (1) the weapon must be made more effective; and (2) failure must be avoided, since, for example, the failure to hit a homing enemy missile can have dire consequences.

The forces along a long trajectory changing the course of a projectile are many and can vary rapidly. If the weapon is controlled at the firing base, where sophisticated control instrumentation is available, it can be more accurately directed than by an on-board instrument. On board the weapon, sensors designed to feedback the target acquisition information are still needed. This could include a TV camera which can provide direct visual feedback to the operations crew at the control center.

The requirement for missile and torpedo guidance has been recognized for some time. Wire-guided weapons of this type are in existence. Since the wire has limited information capacity, the control functions exercised are relatively simple, and no TV viewing capability can be installed. The advent of optical fiber transmission makes it possible to have fiber-guided weapons with on-board TV cameras. This enables the weapon to be guided to regions where direct line-of-sight vision is obscured by the terrain. The electromagnetic interference freedom of using optical fiber is an additional benefit, as it effectively prevents electronic jamming.

DESIGN CONSIDERATIONS

The rapid acceleration involved in the firing of a missile exerts considerable force on the guide wire. The force exerted is dependent on the mass per unit length of the fiber, as well as the rate of acceleration. It is necessary to design a fiber which is small and strong and to wind it in a payout housing (cannister) which allows rapid payout with minimum increase of force on the fiber. The cannister must be of a size convenient for mounting on the weapon and have enough room to enable the fiber to cover a certain range. In general, the longer the range, the greater the demand on the fiber characteristics, since the fiber must maintain high strength over the entire length of the range and must have low transmission loss to allow satisfactory signal transmission.

Cable Design

A reinforced single-fiber cable using a fiber of an overall diameter smaller than 125 μm is a possible design. The fiber is proof-tested to about 2 percent strain so that it can be coiled to a tighter radius and still have an assured long service life. A typical residual strain of 0.5 percent is permissible. The fiber loss is kept as low as practical, but with the numerical aperture kept sufficiently high to prevent microbending loss increases. A range of 5 to 10 km can be met readily. In order to facilitate payout, a single fiber is preferred. Two-way transmission along a single fiber becomes necessary. This can be done by using the fiber in a bidirectional fashion at two wavelengths. This means that at these wavelengths the transmission loss must be sufficiently low to allow the power budget to be met.

Bidirectional Dichroic Coupler

A bidirectional coupler allowing two wavelengths of operation is called a *bidirectional dichroic coupler*. In the missile-guidance application the information from the missile to the control has a wide bandwidth, but the information from control center to the missile is usually at a low bit rate. This means that the relative power of the transmitter and the sensitivity of the receiver can be appropriately adjusted to make the isolation requirements of the coupler less stringent. Furthermore, the broadband information can be handled at the wavelength corresponding to the lower transmission loss.

The dichroic coupler should utilize wavelength spacings sufficiently removed to avoid near-end crosstalk due to residual radiation at wavelengths within the passband of the receiver filter. The wavelength spacing should not be too far apart, such that one wavelength is the harmonic wavelength of the other, thus causing the dichroic filter design to become more difficult.

Transmitter-Receiver Designs

The large-bandwidth transmitter is often required to produce a significant amount of power output in order to allow the power budget of the large-bandwidth channel to be attained. If the TV camera is a 1-MHz signal, the transmitted power is 0 dBm with the use of intensity modulation, and the receiver noise level is at -50 dBm, a 35-dB SNR reception allows a 15-dB margin. Since this is a one-shot affair and all connections can be made by splicing, the entire margin can be assumed to be available for fiber loss. For fiber loss at 3 dB/km, the range is 5 km, where at 1.5 dB/km, the range is 10 km. The low data rate channel has at least 15 dB more margin, since the signal is digital and since the data rate is 100 times smaller. The receiver sensitivity and transmitter power can be adjusted to offset the change in the loss of the fiber and the coupler.

Torpedo Guidance

Torpedo guidance differs from missile guidance in the rate of acceleration, the final velocity, the range, and the environment. It is possible to use a two-fiber cable which is designed at a specific buoyancy. The cable usually needs to incorporate additional strength-rendering materials.

Tethered Vehicles

Instrument vessels and vehicles are sometimes tethered and towed by fixed and mobile data-processing centers, respectively. To allow a higher rate of data transfer, the transmission line incorporated in the cable should have a larger bandwidth capability. The incorporation of optical fiber in such cables provides the large-bandwidth transmission capability. The high strength and small size of the fiber can result in a lower cable cost.

DESIGN CONSIDERATIONS

The only distinct aspect of this design is the cable. For the incorporation of optical fiber into a tow or tether cable, the fiber is required to have good strength and be packaged in such a way that its loss will not be appreciably increased through bending caused by the forces within the cable.

Distribution Systems[5,6]

The distribution of multiservice to a large number of users or subscribers is a scenario of the future communications network. It is an extension of the distribution of a single service such as the telephone service. The local distribution requirements vary according to the population spread in a given area. In the metropolitan area, large cross-sectional cables are needed to reach a concentration of subscribers in high-density dwelling units such as high-rise apartments. In the suburban area, subscribers tend to spread somewhat. Cables designed to cater to a density of 20 subscribers per square kilometer is fairly common. In the rural area the subscriber density drops to less than 20 subscribers per square kilometer. Long cables are, therefore, needed for each subscriber. The economics of the subscriber loop plant, as the equipment and cable associated with subscriber distributions are called, is dependent on the cost of providing the service and the revenue such a service can generate. Obviously, the rural regions present a dilemma if a uniform subscription rate is imposed for all subscribers.

The use of optical fiber systems in a distributed network offers wideband delivery capabilities to each subscriber premise, but the cost can be prohibitively high unless it can increase the revenue base accordingly. Furthermore, the actual method of implementation and the way services are integrated can greatly affect the total operation cost and revenue potential.

Wired Office

The office of the future is designed for an increased usage of data-processing equipment to provide stored reference data for the day-to-day operation of the business; to have analytical capabilities; to assess stored information at information banks, to do word processing; to distribute documents locally and remotely; and to communicate in real time with other offices, with telephone, data, and television in a multiply connected fashion. In this environment optical fiber is expected to enable more optimized networks to be developed. It is to be noted that with low data signals, the use of a wideband transmission cable is not entirely necessary unless a high-speed package switching mode offers a low-cost solution.

Wired City

The wired city is an extension of a wired office. It is a distribution system in which a multiplicity of services are distributed to a large number of subscribers over an extended region. The range of services may be larger than that encountered in the wired office. For example, the types of stored information and the number of TV channels for selectable accessibility can be very large. The mode of usage of the different types of services is likely to depend on the social habits of the people. The proliferation of wired cities could result in a major structural change in society.

DESIGN CONSIDERATIONS

In order to provide a comprehensive multiservice network, the services must be divided into several levels and be introduced in an evolutional manner. A logical division of services is telephone, data, and video. Switches for these three basic services should be provided separately.

If all these services are required, an integrated switchable service is likely to be the most cost-effective solution. The number of optical fiber lines to each subscriber is a minimum of one for a full bidirectional two-way service. The provision of two or three fibers to each subscriber can be advantageous for increasing design flexibility and reliability. When a large area is to be covered, the distribution system must have at least two tiers: the central office and then the nodes where a certain amount of circuit concentration can be implemented. Figures 8-1 through 8-3 indicate a possible scheme. Figure 8-1 shows how the video (14

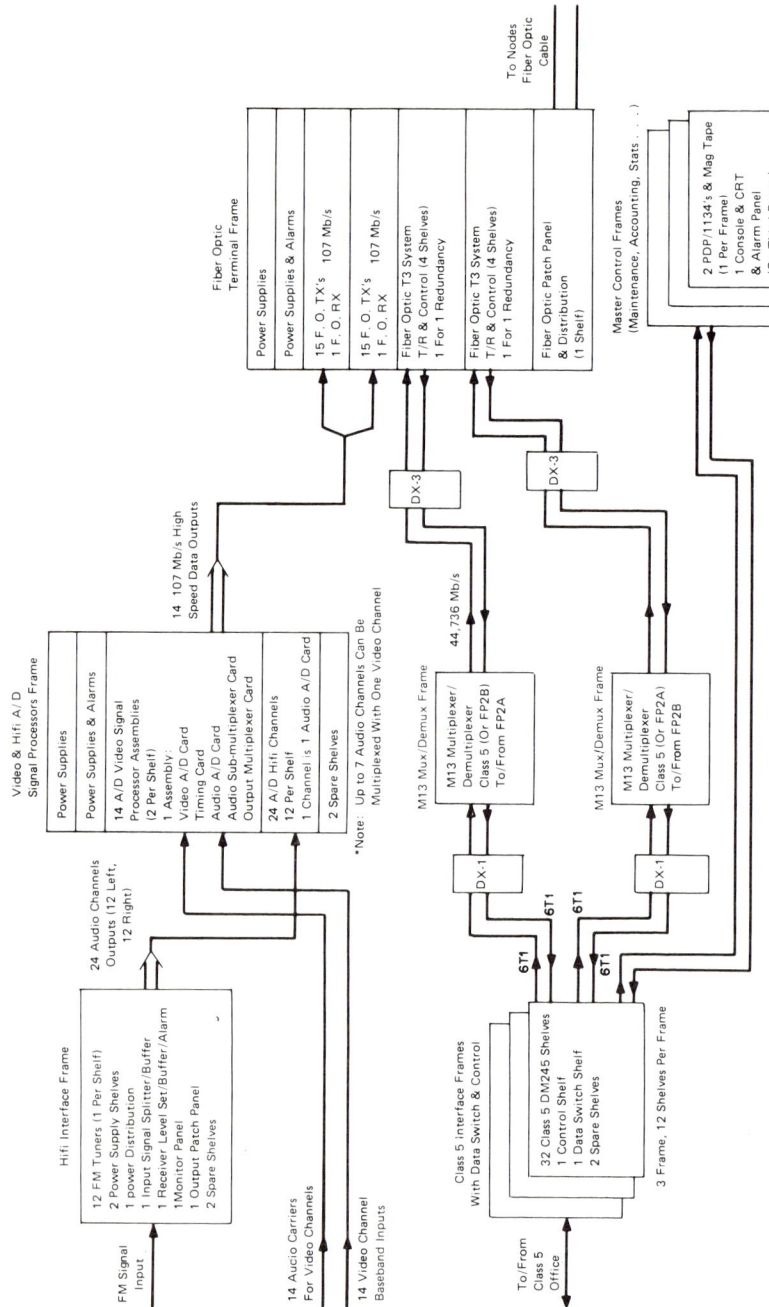

Fig. 8-1 Central office block diagram.

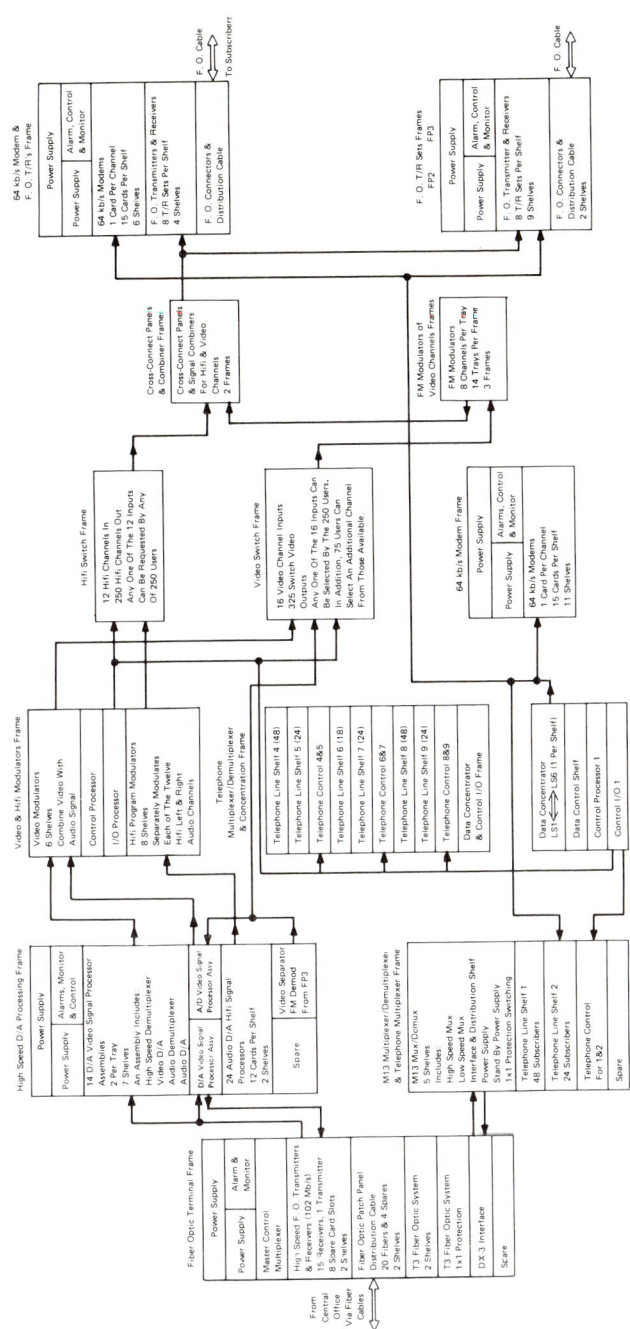

Fig. 8-2 Nodal office block diagram.

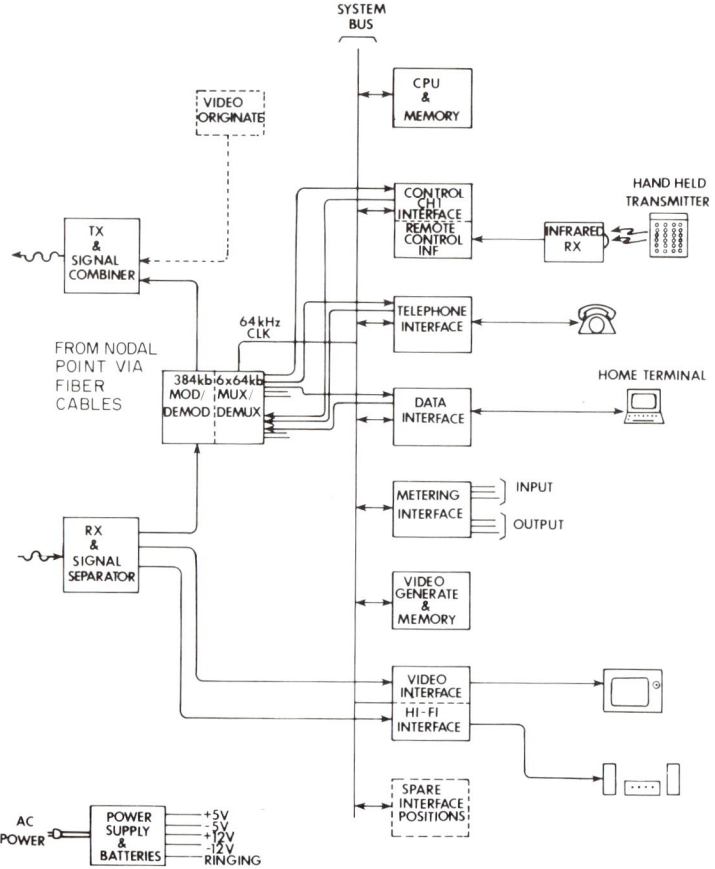

Fig. 8-3 Subscriber home block diagram.

channels), hi-fi (12 channels), and telephones at the central office are assembled and readied for onward transmission to the nodes. Figure 8-2 illustrates the scheme for circuit concentration for telephone and data traffic and for TV distribution via video and hi-fi switches. Figure 8-3 shows how a subscriber home is served.

The subscriber is to receive at least two TV channels simultaneously from a choice of any number of available TV vendor channels. The number of TV channels available and the charging plan depend on the type of service. Presumably, a service catering for the minority will be at a higher cost and may have a higher intrinsic value. If a videophone service is included, then the bandwidth for TV transmission to the home must be at least sufficient for three TV channels. The TV channels and the videophone channel will share the same video switch with the TV channels wired differently to the videophone to cater to the different mode of usage. The videophone circuit will first be processed at a concentrator.

The other services such as FM radio, telephone, and data can be handled as a separate band of information for downward distribution to a subscriber, together with the two TV channels and one video channel in an analog modulation format.

The quality of TV with the use of analog techniques is only just acceptable. To improve TV quality and enable distribution distances to be increased, the TV signal should be digitized. With two-to-one and eventually four-to-one compression, two or four TVs can be accommodated in a 100-Mb/s bandwidth. More TV channels per subscriber, easier switching, and low mass-production cost are some of the other advantages. A full integrated network can also be envisaged more easily.

The upstream channel carries the videophone, the telephone, and data services. This channel is less demanding in quality and hence can be implemented more readily. The data upstream is for command and control in order to achieve a two-way communication. It can also be used to provide services such as fire and burglar alarms, as well as the supply of electricity consumption data for billing and electrical load control purposes. The telephone can be used as the key to set up the services, since these functions are invariably available at the switching central office of the present network.

Distribution Cable

A cable with a large number of fibers arranged in subbundles of different sizes is required. For ease of laying, the subbundles should be stranded in an oscillatory way into a bigger bundle. This is necessary to permit branching and to remain flexible. The use of wire boxes where cables are to be spliced for branching can be adopted with similar benefit as in the case of the copper distribution cable. Colorband coding can also be adopted for ease of identification.

Advantages

A wired city with broadband distribution capabilities will create a community where information management is efficient and energy usage is balanced. Out of this pioneering community will come the blueprint for understanding of future social needs. These clues are needed for planning toward the future.

A broadband distribution with switchable TV is to make TV serve different sectors of population in small groups instead of en masse. This would promote better programming and more meaningful programming. There will be many new service-oriented industries arising from the wired-city concepts. Its introduction will also allow the broadband technology to be exploited to its fullest.

Cable TV Distribution

Cable TV distribution is the major cost component of a cable TV system. The cable TV industry must invest a very large sum of money to wire up a city. The current method presents two major problems: maintenance and the difficulty of introducing extra channels for subscribers paying additional subscription. Cable TV operators will use switchable TV if the distribution cost is not excessive.

Distribution of switchable TV by optical fiber is expensive and cannot be envisaged for TV distribution only. In fact, the basic cost probably is about three times higher. In some ways this is not really excessive, but the perceived value may not be sufficient to make a customer pay three times the premium for this new service.

DESIGN CONSIDERATIONS

An optical fiber analog TV distribution system can be envisaged with the use of a high-power laser and a high-linearity external modulator. The TV signal can then be carried on an analog modulation basis with perhaps 12 channels per laser wavelength. A double-tier star-distribution network as shown in Fig. 8-4 indicates that such a distribution is possible. The only advantages, if the network is for distribution of TV only, are the improved reliability and the lowered maintenance cost.

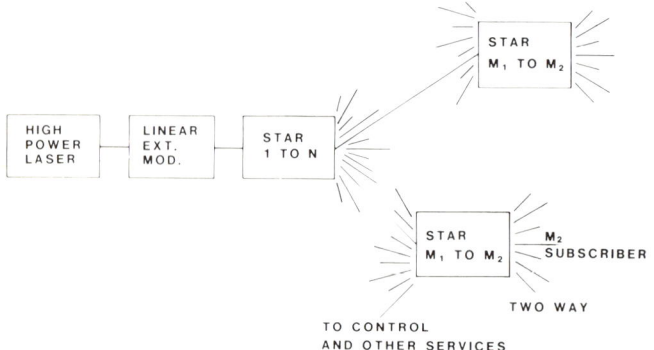

Fig. 8-4 A double-tier star-distribution scheme.

Sensor Systems[7]

Basic to the single-mode fiber transmission property is its ability to transmit a single mode over long distances. This means that the optical phase information is retained after transmission. This property can be exploited for sensor applications. For temperature or pressure sensing, the small variation of the effective fiber length due to temperature or pressure is made to produce measurable phase changes. If a long fiber is used, a minute change per unit length will be greatly magnified over the entire long length to give a large phase shift. This phase shift, if measured by comparison with a reference, will be a measure of the temperature or pressure change. For rotation sensing, the change in the light velocity in a fiber coil of many turns produced by the rotation is measured. If the fiber coil is made sufficiently sensitive to respond to the rotation of the earth, it becomes a gyroscope and can be used as a navigational aid. For magnetic field sensing, the fiber length must change with the magnetic field. This is achieved if magnetostrictive material is used to coat the fiber. A measure of the change of fiber length gives the strength of the magnetic field.

ACOUSTIC SENSOR

When an acoustic wave impinges on a fiber, the refractive index of the fiber material is modulated by the acoustic wave such that the light wave traveling along the fiber will undergo phase changes. By detecting the phase change, the acoustic variations can be measured. Thus the fiber becomes an acoustic sensor.

For measurement of phase changes, the fiber must retain phase information. The first major problem is to design a fiber with a unique polarization even after bending. This is difficult for two reasons: (1) a preferred polarization direction must be defined; and (2) this polarization direction must be registered such that it can be recognized external to the fiber. Methods such as the use of an elliptical fiber of high NA and the use of strain birefringence are available to maintain the designed polarization. It is possible to use a fiber carrying the TE_{01} mode.[8] Unlike the HE_{11} mode, this mode has complete circular symmetry so that it has a unique phase velocity.

The application of fiber as an acoustic sensor is just one more example of the use of optical fiber as a sensing element. Many potential usages are yet to be explored. It is important to point out that optical phase change can be produced so readily that the sensitivity of fiber as a sensor is very high, but a high-sensitivity element must be carefully treated. For example, acoustic sensors must be compensated for temperature changes, since temperature variations will produce large phase changes.

References

1. E. E. Basch, R. A. Beaudette, and H. A. Larnes, "Optical Transmission for Interoffice Trunks," *Transact. IEEE* **Com-26**(7):1007–1014 (1978).
2. F. Aoki and Nabeshima, "Optical Fiber Communications for Electric Power Companies in Japan," *Proc. IEEE,* **68**(10):1280–1285 (1980).
3. C. D. Anderson, R. F. Gleason, P. T. Hutchison, and P. K. Runge, "An Undersea Communication System Using Fiber-guide Cables," *Proc. IEEE* **68**(10):1299–1303 (1980).
4. M. K. Barnoski, "Fiber Systems for the Military Environment," *Proc. IEEE* **68**(10):1315–1320 (1980).
5. K. Y. Chang, "Fiberguide Systems in the Subscriber Loop," *Proc. IEEE* **68**(10):1291–1299 (1980).
6. W. E. Herold and H. Ohnsorge, "Optical Fiber System With Distributed Access," *Proc. IEEE* **68**(10):1309–1315 (1980).
7. B. Culshaw, D. E. N. Davies, and S. A. Kingsley, "Fiber Optic Strain, Pressure, and Temperature Sensors," *Proc. 4th ECOC, Genoa, Italy:* 115–126 (September 1978).
8. H. M. Barlow, "Optical Fiber Transmission in the TE01 Mode," *J. Phys., D, Appl. Phys.* **13**:369–375 (1980).

Chapter

Anatomy of a Design

Introduction

Every system is required to fulfill a set of specifications. A successful design achieves the target specifications within many constraints set by technological performance, cost, availability, background experience, and other factors. It is easy to envisage that many design variations will exist. It is also easy to see that the design process is neither unique nor always explicit. However, the broader design steps for a fiber optic transmission link are traceable, and the rationale for some of the detailed aspects of the design can be readily discussed. In this chapter a design of a fiber optic transmission link for a mixed video, audio, and data signal is to be analyzed.

SYSTEM SPECIFICATION

System specifications usually can be divided into system-level specifications and component-level specifications. For major systems to be equipped, furnished, and installed by a contractor, the system-level specification, giving overall requirements only, may be specified. The method of implementation and the choice of system components are left to the vendor to specify. However, for reasons of ease of maintenance and operational efficiency, the purchaser often calls for standardizing equipment practice and service procedures, thus leading to specifying separate subsystems and modules. Even at the individual device

160   Chapter Nine

level, specification of performance is often called for to enable the user to have the option of making substitutional replacement of these devices.

A system specification may be generated by the user or by the vendor. In the former case the user perceives its need and generates its system specification which the user expects the vendors to meet at an affordable price. In the latter case the vendor perceives the market needs and specifies its system products accordingly. In both cases the specification usually is a balance of what is technically achievable and what can be met at an affordable price. Obviously, considerable knowledge of design is needed to guide the specification generation.

BACKGROUND NARRATIVE OF THE SYSTEM TO BE DESIGNED

Two analog video channels with their associated sound, three analog telephone channels, and one digital 64-kb/s data channel are required to be transmitted from a central office to a remote switch station from which they are to be distributed as a single frequency-division multiplex (FDM) combined signal to one of many subscriber terminals. The signals are required to be transported at an acceptable quality over a maximum length of 10 km with no intermediate repeaters between the central office and the remote switching station, and over a maximum length of 2 km between the remote switching station and the subscriber premises. This portion of the route is also required to carry a low-bit-rate signal from the subscriber to the remote switch station.

These signals and transmission requirements are likely to be encountered in an information-distribution system involving switchable video and are envisaged to be needed in an information-based society.

DESIGN SPECIFICATION

Signals to be transmitted are as follows:

    Two, baseband NTSC standard color TV channel with sound
    Three, 300- to 3400-Hz telephone channels
    One, 64-kb/s digital binary non-return-to-zero (NRZ) format data channel

The signal quality at the remote switch station after transmission is as follows:

1. TV: (a) SNR_{vhf} or SNR_{BB}—50 db; (b) differential gain—3 percent; (c) differential phase—2°; (d) peak-to-peak periodic noise——60 dB;

(e) frequency response—±0.3 dB; (f) sync compression—1 percent; (g) field time waveform distribution—0.5 percent.
2. Telephone: standard 64-kb/s PCM channel performance.
3. Data: 1 in 10^8 BER.

The signal quality at a subscriber terminal is as follows:

1. TV: (a) SNR_{vhf} or SNR_{BB}—40 dB; (b) differential gain—10 percent; (c) differential phase—5°; (d) peak-to-peak periodic noise—50 dB; (e) frequency response—±1 dB; (f) sync compression—3 percent; (g) field time waveform distribution—3 percent.
2. Telephone: standard 64-kb/s PCM channel performance.
3. Data: 1 in 10^8 BER.

It is to be noted that the acceptability of signal quality is subjective and varies from person to person. The quality specification used above is intended for this design discussion and does not imply an accepted standard of customer acceptability.

TOP-LEVEL SYSTEM DESIGN CONSIDERATIONS

The system is designed to have optical fiber transmission between the central office and the remote switch station and between the switch station and the subscriber. The two sections need different transmission schemes, since the technical and economic requirements are different. In the former, to be designated as section 1, the span length of the link is 10 km, and in the latter, to be designated as section 2, the span length of the link is 2 km, but it is required to be designed for low cost.

Obviously, if the different channels are multiplexed into a single signal and transmitted along a single fiber, the fiber transmission line cost is a minimum. Whether this is possible while the signal quality can still be maintained is the crucial consideration in the design. Furthermore, the cost of multiplexing and demultiplexing must be considered, as it may offset the transmission line cost advantage.

On a FDM basis, the TV signals and the other signals can be placed on subcarriers and assembled into a single signal of certain bandwidth. As will be seen later, a convenient way is to assemble the two TV channels at standard very high frequency (VHF) channel frequencies and digitalize the telephone channels, time-multiplex them with the data channel, and then place them together on a subcarrier near zero frequency. This offers the least intermodulation interference between the different channels of information and also allows the TV signal to be readily received by a standard TV receiver. This entire band of information is to be transmitted with appropriate distortion levels and SNRs as defined by the specification.

On a time-division multiplex (TDM) basis the TV signals, the telephone, and the data can all be digitalized and assembled into a single-bit stream. As will be shown later, a 200-Mb/s pulse rate is needed if each TV channel is encoded into a digital signal of about 100 Mb/s. However, through the use of signal processing techniques, the bit rate for a single TV channel can be reduced by a factor of at least 2.

If the channels are to be handled separately, the TV signals can be handled conveniently as baseband analog signals or as PCM encoded 100-Mb/s signals. The three telephone and one data channels are more readily handled as a 4 × 64-kb/s channel, since digitalizing a telephone channel into a 64-kb/s PCM signal is a standard practice which can be readily achieved by using a mass-produced circuit known as a *single-channel codec*.

SECTION 1 DESIGN CONSIDERATIONS

Analog Signal Transmission Considerations

The signal is to be transmitted over 10 km. First, we consider the link budget. Let an LED transmitter be designed to give a power into fiber at −10 dBm with a linearity sufficient to accommodate a composite signal of ∽100-MHz bandwidth. The receiver sensitivity is designed to be −20 dBm for an SNR of 50 dB. This means that there is only 10 dB of loss permitted between the transmitter and the receiver. This is hardly enough for allocation to each loss component. If a laser transmitter is used with a 0-dBm power coupled to the fiber, a margin of 20 dB is available for allocation to fiber loss, splice loss, and time-dependent degradations. The use of laser, however, gives rise to various noise contributions, thus limiting the SNR attainable. Since the specification calls for >50 dB SNR, analog transmission is ruled out.

Digital Signal Transmission Consideration

The link budget is considerably different for digital signals, since an SNR of only about 20 dB is required to meet a BER specification of 1 error in 10^9. Thus, using an LED, the margin is 40 dB and using a laser, 50 dB. Fiber loss of ≤3 dB/km can be used. The bandwidth for a single TV channel is ∽100 Mb/s. This calls for a fiber with a dispersion of <10 ns over 10 km while for two TV channels to be simultaneously transmitted, the dispersion is <5 ns over the 10 km. This loss and dispersion are both readily met with a graded-index fiber. It is to be noted that the use of LEDs may not allow >100-Mb/s operation. A digital transmission for section 1 is, therefore, more appropriate. Whether the signal should be a single-bit stream of 200 Mb/s or two separate bit streams of 100 Mb/s to be transmitted along one fiber by wavelength multiplexing or along two fibers is the design alternative.

SECTION 2 DESIGN CONSIDERATIONS

Analog Signal Transmission Considerations

The signal is to be transmitted over 2 km, and the signal quality requirement is much less stringent than that called for in section 1 of the system. An SNR of 40 dB is attainable with a laser link or even with an LED link.

There are two advantages in using the analog format: (1) the multiplexing and demultiplexing costs are low; and (2) the signal processing cost associated with TV reception is a standard product. These are important advantages when cost is a serious issue.

Digital Signal Transmission Considerations

Apart from transmission distance reduction, the discussion of section 1 design applies here. In fact, since the distance is shorter, the transmission is much easier, especially relative to the dispersion needs.

However, the signal processing cost is a critical issue. In principle, the introduction of appropriately designed mass-produced circuits could reduce the cost sufficiently.

RECOMMENDED DESIGN

A digital link for section 1 and an analog link for section 2 are the recommended designs after this top-level system design consideration. The graded-index fiber operating at 0.85 μm appears to be adequate. The detailed specification and design are discussed at the system, equipment, and selected component levels.

Section 1 Link Design

SECTION 1 LINK SYSTEM SPECIFICATION

A digital optical fiber link is to be designed to carry a single PCM National Television Standards Committee (NTSC) TV signal at 89 Mb/s, with the following specifications:

- Link span length—10 km at 100 Mb/s
- Laser transmitter output into 50-μm-core graded-index fiber—0 dBm at 0.85 μm
- Receiver sensitivity—−50 dBm at 100 Mb/s
- Fiber loss—3 dB/km
- Fiber dispersion—10 ns over 10 km

LINK BUDGET

The link budget is presented in the following table.

Transmitter output	0 dBm
Receiver sensitivity	−50 dBm for SNR of 20 dB
Margin	50 dB
Fiber loss 3 dB/km, 10 km	30 dB
Total splice loss, 0.3 dB × 10	3 dB
Total connector loss, 1 dB × 4	4 dB
Detector coupling loss	1 dB
Headroom for temperature range and aging and future splices	9 dB
Total attenuation	47 dB
Excess power margin	3 dB
Fiber dispersion	10 ns over 10 km

INTERPRETATION OF SECTION 1 FIBER OPTIC LINK SPECIFICATION

To meet the specification, the video signal is digitized to eight times the third harmonic of the color subcarrier frequency of 10.738635 MHz or ∼86 Mb/s. This 8-bit PCM linear encoding achieves the SNR, differential phase and amplitude, and other permissible distortion levels as stated in the specification. This bit stream is forwarded to the digital transmitter input.

The laser transmitter is designed to output the optical signal at 87 Mb/s with an average optical output power of about +3 dBm such that when coupled to a 50-μm-core graded-index fiber of 0.2 NA, the launched power is 0 dBm.

The transmission loss and dispersion along the fiber cable is designed to be <30 dB and <10 ns over the 10 km. The cable is designed to be installed in a duct and is to be installed in a 1-km length. The cables are to be fusion-spliced with splice placed in sealed housings. Splice loss is to be <0.3 dB per splice. The cable is terminated with connectors at a patch panel and connection to the transmitter and receiver equipment optical ports via patch cords terminated in optical connectors with a loss of <1 dB per connection.

The receiver is designed to have a sensitivity to achieve a BER of <1 in 10^9 when the input power is at -50 dBm average power at 87 Mb/s. The receiver uses an APD and a transimpedance design to achieve the performance goals.

ANALOG-TO-DIGITAL CONVERSION

The video signal is assumed to be at baseband while the associated audio signal is also assumed to be at baseband. The two signals are to be individually converted from analog to digital bit stream and transmitted.

For the video signal, the analog-to-digital (A/D) converter must accommodate a range of signal levels. If the video sync tip level is locked to a dc level, the usage of the A/D converter's voltage range can be adjusted to permit a wide dynamic range. Accordingly, a sync clamp circuit is to be used. Following the clamp, the signal is processed through a video A/D converter such as a TRW model TDC1007J monolithic large-scale integration (LSI) chip or a Computer Labs MATV-0816 A/D converter module which uses hybrid circuit technology. The former is an 8-bit parallel A/D converter capable of accurately sampling, without an external sample and hold circuit, an input signal with frequency component up to 7 MHz. The latter contains an input track and hold amplifier, followed by a 8-bit parallel A/D converter. Both are capable of delivering a differential phase of 0.5° and a differential gain of 1.5 percent when measured for an NTSC ramp modulation and unlocked sampling.

For the audio signal, the A/D conversion digitizes the signal into a 14-bit parallel word with a 75-μs preemphasis at 767 kb/s. This rate is expanded by a factor of 2 and stuffed to 1.5 Mb/s and transmitted as a T1 carrier. This form of PCM encoding of an audio channel achieves a signal of hi-fi quality with the following levels of imperfections:

Total harmonic distortion—<0.1% 1 kHz at +18 dBm
Unweighted noise (20 to 15,000 Hz)—<−70 dBm
Periodic noise (<20 kHz)—<−73 dBm

DIGITAL VIDEO TRANSMITTER

The video transmitter is designed to operate at about 100 Mb/s. A narrow-stripe GaAlAs laser, with low threshold current and an emission area suitable for low-loss coupling into a 50-μm graded-index

166 Chapter Nine

fiber, is to be used. A low-threshold current means a low transmitter power consumption and is an advantage. However, the choice of the laser is often determined by power output and coupling considerations. Furthermore, the problem of multipath interference requires the laser to be designed for relatively broad spectral width.

A possible transmitter circuit is as shown in Fig. 9-1. The circuit consists of a drive circuit for the laser, the provision of a laser prebias point, and an optical output power control through the use of a circuit which detects the output power fluctuation of the laser and corrects this through a feedback loop. The laser output can be monitored by tapping a portion of the output at a beam splitter or using the back surface radiation from the laser. In both cases a PIN photodiode is used for detection. The former method simplifies the laser casing design, since the LED is separately mounted, but could aggravate multipath fading effects and increasing laser noise. However, the latter method could encounter difficulties if the laser front and back face output do not track exactly. It also requires the laser to be mounted with access to both front and back emission regions. This complicates the mechanical design of both the laser casing and/or the fixture on the circuit board.

An aged laser monitor is fitted as a means of performing preventive maintenance. It derives the aging information by measuring the laser drive current.

The optical output from the laser transmitter is to be conveniently and efficiently coupled to the optical fiber transmission line. This calls for the laser to be designed with a package which has an output port designed to allow efficient coupling to the particular fiber transmission

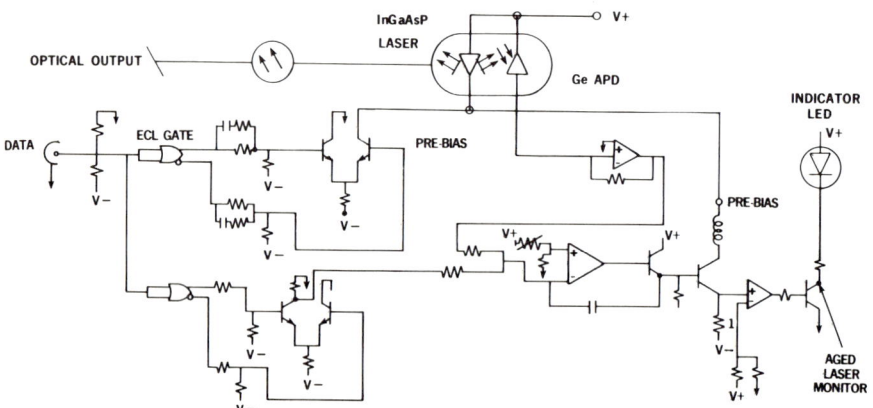

Fig. 9-1 Transmitter functional schematic.

line. It is to be noted that the reflected signal at the connector could cause laser instability and model noise. An isolator, which offers low transmission loss in the transmit direction and high loss to the reflected signal, is required.

The transmitter design determines the input signal level, the electrical power, the heat-dissipation arrangement, the available peak and average power, and the expected coupling loss to the fiber transmission line. The physical design of the transmitter is constrained by the equipment practice. For example, all inputs and outputs in certain equipment practice are at the back of the equipment rack, or all functional units are on printed circuit board cards. In this case the optical connector must be designed to allow backplane plugin. As another example, electrical signal ports handling high signals should not be allowed to radiate excessively in order to prevent the occurrence of crosstalk.

DIGITAL AUDIO TRANSMITTER

The audio signal digitized into a 1.544-Mb/s T1 rate bit stream is fed to a T1 digital transmitter. This same type of transmitter is to be used for transmitting the telephone and data signals.

One design approach is illustrated here. This is aimed at simplifying the circuit design by working at a higher than minimum bit rate. This design approach is perfectly acceptable for optical fiber systems, since

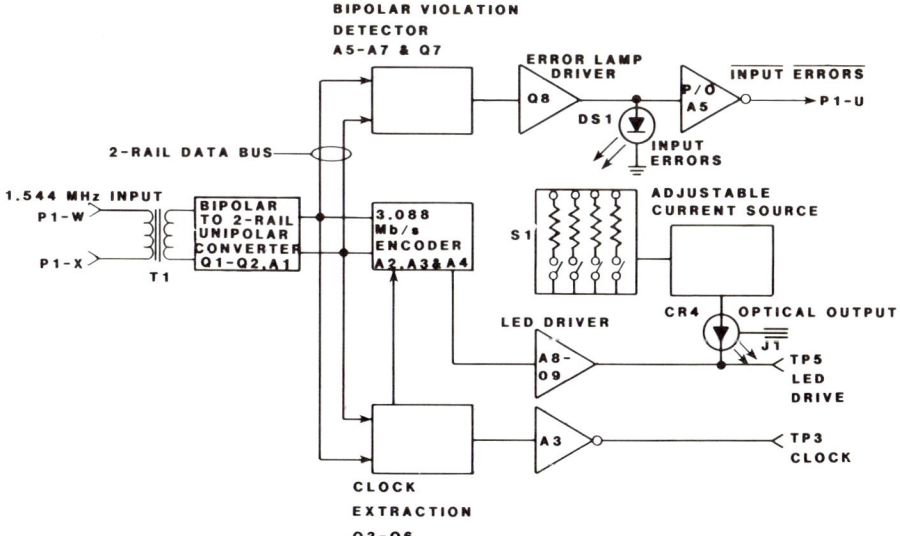

Fig. 9-2 A 1.544-Mb/s T1 transmitter schematic.

the transmission capacity of the fiber is capable of handling the higher bit rates without cost disadvantages. Moreover, the use of a higher than minimum bit rate could achieve error supervision and signal scrambling simply and effectively.

The transmitter schematic is as shown in Fig. 9-2. The incoming 1.544 bipolar data, from standard T1 multiplexor output, is converted to two-rail unipolar format and then to a 3.088-Mb/s NRZ signal. Since the bit rate is low and the power budget is adequate, an LED source is used. This provides high reliability, larger working temperature range, and simpler drive circuit requirement. An output of 50 μW or -13 dBm is launched into a 50-μm fiber. This gives a margin of about >45 dB at 3 Mb/s with an APD detector. The power budget then becomes:

Transmitter power	−13 dBm
Received power	−58 dBm
Margin	45 dB
Fiber loss	30 dB
Splice loss	3 dB
Connector loss	4 dB
Detector coupling loss	1 dB
Headroom	4 dB
Total	42 dB
Excess power margin	3 dB

The headroom in this case is taken to be smaller than that needed for a laser transmitter, since the temperature and aging of LED is less.

The circuit also included a bipolar violation detector which offers an alarm facility as well as providing an output for remote violation counting. A clock extraction circuit is needed to assure synchronization of the incoming 1.544-Mb/s signal and the 3.088-Mb/s signal.

An adjustable current-regulating circuit with a 12-dB adjustment is incorporated to cater to LED variations and is valuable as an installation and troubleshooting aid. Several test points are provided. These are a ground test point, a clock test point to check the output of the clock extraction circuit, and a test point for checking the 1.544-Mb/s input signal. The dc power required for the LED drive circuit is 5 V, 0.3 A. The signal input is typically 3 V ±10 percent into a 100-Ω balanced load.

This mode of transmitting the audio channel is unattractive economically, since the fiber cost is relatively high and is totally underutilized. An alternative way is to transmit the audio channel as part of a larger band of signals. For example, this audio channel, together with the telephone and data signals at 64-kb/s per channel from many subscribers, is combined into a number of T1 bit streams and multiplexed to T2 or even T3 rate for transmission along the same fiber.

DESIGN OF THE OPTICAL FIBER CABLE FOR SECTION 1 TRANSMISSION

It is assumed that the cable needed for this section is to be laid in preconstructed and partially filled wire ducts. The manhole spacing is around 0.5 km, and that access is limited. The temperature range is from $-20°C$ to $+55°C$, and the relative humidity can reach 100 percent. The duct may be momentarily water-filled, but no rodent infestation is expected. Splicing can be executed at all manholes, but no electricity supply is available. The pulling tension through the 0.5-km section is not expected to exceed 100 kg for a cable of diameter less than 1 cm.

The requirements stated above influence the design of the cable to a certain extent. Together, they call for a highly flexible cable with no metallic sheath but with a jacket material capable of withstanding duct pulling and a filled cable to avoid water ingress. The number of fibers in the cable is four; one for transmitting the video, one for the audio, and two for the duplex telephone and data channels. No power-carrying conductors or order wire are called for. A viable design is a centrally located fiber bundle with strength-rendering material placed around it and is finally jacketed with a protective polymer coating.

FIBER UNIT DESIGN

Three principles are followed in this design:

1. The fibers are to be stranded for flexibility. The choice of the pitch influences the stranding machine design, the payout tension control accuracy, and the residual stress and twist magnitudes. A 5-cm pitch is quite easily achieved and provides a highly flexible design. Structural and material uniformities prevent increase in fiber loss due to the presence of unbalanced forces.
2. To minimize the effect of differential expansion coefficients of plastic and glass, which becomes a problem at low temperatures, the plastic materials used for the fiber jacket should have a low transition temperature, and an oversized fiber is used as a center member to increase the glass-to-plastic ratio, which tends to prevent the plastic shrinkage from causing severe microbends.
3. A void-filling compound is to be used to prevent significant water

accumulation within the fiber bundle structure. Since the cable is to be used at subzero temperatures and since the cable materials are not impervious to water, an alternative is to design the cable for gas pressurization. However, this alternative design would increase the cable diameter as well as the operating cost.

FIBER LOSS AND DISPERSION

A well-designed fiber with protective polymer coating can readily have a fiber loss and dispersion of <3 dB/km and <1 ns/km. The incorporation of the fiber into the cable structure gives rise to several issues.

Excess losses due to cabling and temperature variations can result from poor dimensional tolerance, low numerical aperture, nonuniform coating and nonuniform lateral pressure from the cable structure, and uneven strength members. A well-designed fiber cable should minimize the excess loss, so that cabled fiber loss and dispersion are still within the specification.

The effect of concatenation is the effect of joining different fibers together to make up the entire length. In a 10-km cable, fibers may or may not be made in 10-km length. Furthermore, for ease of cable laying, maximum uncut cable length is often less than 10 km. In the present case the maximum cable length for duct installation is stated as 2 km. Thus different fibers are to be spliced together to form the total span length.

The diameter, the concentricity, the ellipticity, the numerical aperture, the fiber profile variations, and the perfection of the splice all contribute to modify the fiber transmission characteristics. The effect of all these factors is to cause total loss to be somewhat larger than the addition of the losses of each individual length and a dispersion somewhat smaller than the addition of the individual components. It is often easier to specify the total loss and dispersion requirement and leave the vendor to figure out the best way it proposes to achieve them. It is important to note that repair and replacement of a portion of the fiber may be needed during the life of a system. Such replacement must not alter appreciably the transmission characteristics. This means that the use of specially chosen fiber design to achieve dispersion compensation, for example, may result in special maintenance requirements.

FIBER STRENGTH AND DURABILITY

The forces involved in pulling the fiber cable through a partially filled duct are due principally to the friction between the cables. If the fiber cable is trapped between the interstices of two other cables, the pulling process may have to be aborted. This is no different from pulling any cable into a partially filled duct. The use of a lubricant and an appropriate leader wire could ensure successful laying of the cable.

The fiber cable is to be designed to withstand the maximum load anticipated during the pulling exercise. To prevent failure during the entire life of a system, the residual stress within such a cable must be sufficiently small to minimize or prevent the occurrence of fatigue failure. The duct-laid cable is not expected to be pulled into the duct more than once, so retention of the high strength during the entire service life is not necessary. This contrasts to a submarine cable, when laying stress is to be encountered at the initial laying as well as when the cable is fished out for repair.

A cable designed for duct installation should be designed with a relatively small cross section, since the cost in a shared duct is proportional to the fractional area the cable occupies.

A design is as shown in Fig. 9-3. This cable has color-coded fibers for identification purposes and uses Kevlar-49 as the strength member. Since Kevlar has high tensile strength, high modulus, large elongation to break, and low density, it results in a small and compact fiber cable. A polyethylene outer jacket combines good mechanical properties at

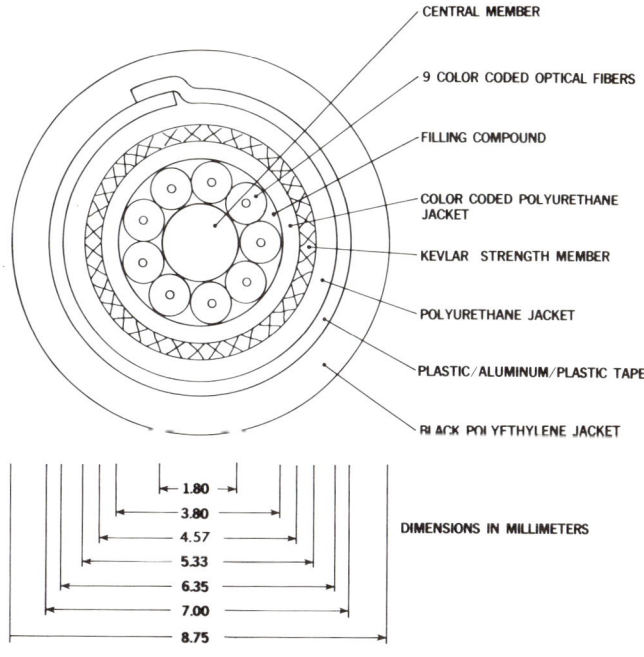

Fig. 9-3 A nine-fiber cable.

172 Chapter Nine

low cost. The cable is designed to have a rated tensile strength of 1000 kg for 1 percent elongation. The maximum tension expected for a 1-km cable with no lubrication is 100 kg. A list of specifications is as follows:

Temperature	
Storage	−50–60°C
Operation	−20–55°C
Number of fibers	4
Fiber attenuation	3 dB/km
Total attenuation over 10 km	<30 dB/km
Fiber dispersion	1 ns·km
Total dispersion over 10 km	<10 ns·km
Fiber core diameter	50 μm
Fiber OD	125 μm
Fiber jacket OD	500 μm
Fiber coating	mechanically strippable
Minimum bending radius	1 m
Impact	0.2 m·kg

SPLICE HOUSING

Cables are joined by splicing. Splice housing is necessary for storage and protection of the spliced fibers. In the telecommunications industry many weatherproof splice housings have been developed for accommodation of copper cables. These are generally adaptable for use with fiber cables. The spliced fibers are separately covered by protective coating and placed within the splice house. Spare lengths of both fiber and cable are also stored within the splice housing, so that resplicing can be repeatedly performed (see Fig. 9-4).

The use of splicing influences the system power or dispersion budgets,

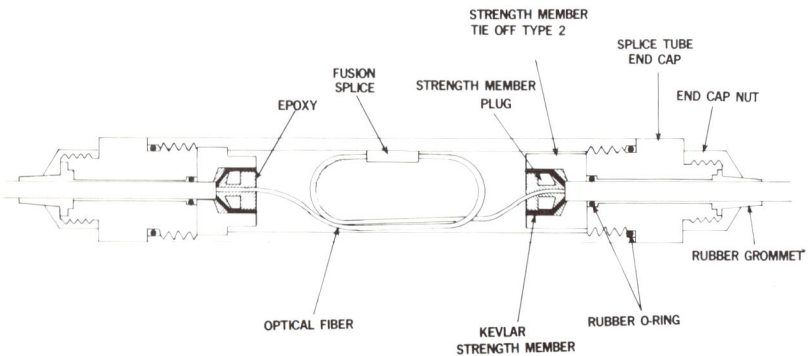

Fig. 9-4 Splice housing.

as discussed earlier. It also imposes a tolerance requirement on the geometric specifications of the fiber, and gives rise to an added material requirement in order to ensure that fibers can be spliced together. Other impacts are in the area of system maintenance and troubleshooting. At the splice housing, splices can be opened for convenient injection of signal for measurement and diagnostic purposes.

VIDEO RECEIVER DESIGN

The video receiver is a high-sensitivity digital receiver using APD as the photodetector. The block schematic of a design is as shown in Fig. 9-5. For improved shielding, the APD and the preamplifier are located within one compartment. The preamplifier consists of a balanced transimpedance amplifier with a cascode implementation to minimize input capacitance. The transimpedance amplifier is designed to have a gain of 4000 and with a flat response beyond 80 MHz. A low-pass filter follows the amplifier to provide equalization and improve BER performance. The signal will then be fed into a limiting amplifier before the time recovery section of the circuit.

At a bit rate of near 100 Mb/s, a phase-locked timing recovery circuit is preferred. An effective design can be achieved offering positive acquisition and adjustable optimal decision threshold setting. The recovered clock and data are fed to the D/A converter at emitter-coupled logic (ECL) levels.

Such a receiver can be designed to have a sensitivity of better than

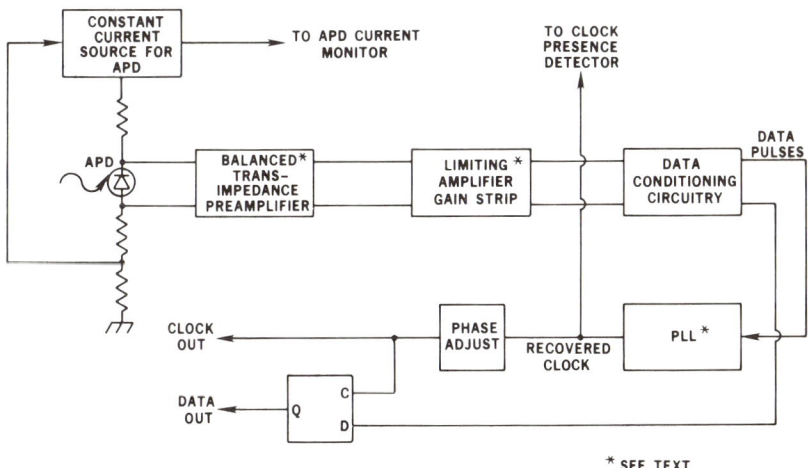

Fig. 9-5 Block diagram of a video receiver.

−50 dBm and will need about 1.5 W of dc power from a 5-V supplier. The APD will need a bias voltage of up to 100 V. Functional indicators for received power and phase-lock condition are usually provided.

VIDEO DIGITAL-TO-ANALOG CONVERSION

The D/A conversion recovers the serial digital bit stream to an analog video signal. The conversion process is illustrated in Fig. 9-6.

The incoming signal is serial-to-parallel-converted into eight parallel bit streams and buffered by line receivers. The ECL levels are then converted to transistor-transistor logic (TTL) in order to match the video D/A converter. The converters are designed to eliminate any time jitter between bits introduced by differential delays between the data lines. The output signal is filtered and buffered before equalization of the sin x/x frequency roll-off induced in the analog signal due to PCM sampling effect.

DIGITAL VIDEO VOICE TO ANALOG CONVERSION

The serial bit stream is rate-reduced, destuffed, and timing-clock-extracted. The data serial bit stream at TTL levels is fed to the standard audio D/A converter. A sample-and-hold circuit is used on the D/A converter output to eliminate conversion glitches. The sample-and-hold circuit is designed with certain acquisition characteristics as a means to equalize distortions.

AUDIO RECEIVER DESIGN

The audio receiver is a 1.544-Mb/s digital receiver. The block schematic is as shown in Fig. 9-7. With the use of an APD detector, −58 dBm sensitivity can readily be achieved for a BER of $<10^{-9}$. This design is similar to the high-bit-rate digital video receiver, except that timing extraction circuits are much simpler.

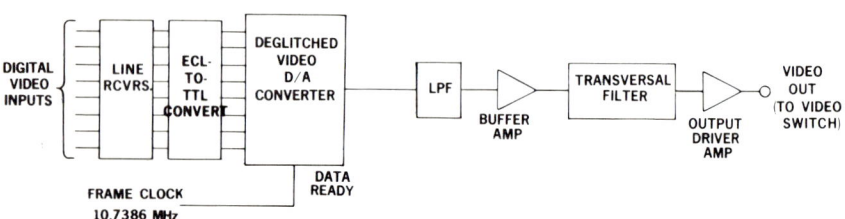

Fig. 9-6 Video D/A conversion card.

Anatomy of a Design 175

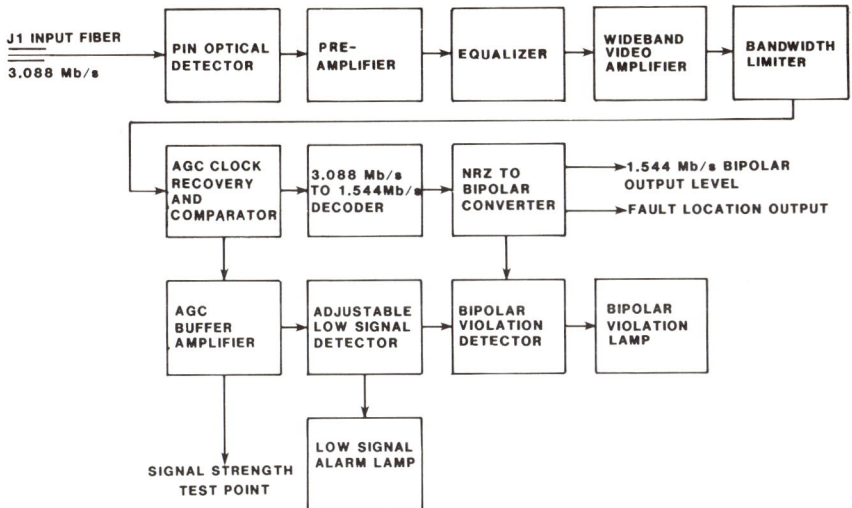

Fig. 9-7 Block diagram of a 1.544-Mb/s optical receiver.

SUMMARIZING SECTION 1 DESIGN DETAILS

Section 1 design details have been described. The equipment consists of the A/D conversion units, the digital transmitters and receivers for video and audio signals, the transmission line, and the D/A conversion units. The design must achieve repeaterless transmission of the video and associated audio signal over a distance of 10 km. It is to be noted that most of the detailed designs presented here are relatively standard electronic circuit designs, even though the constraints set by the electrooptical converting devices are unique. This gives rise to only minor special requirements, such as optical output power control circuit for the laser transmitter and the APD temperature-stabilization circuit.

Section 2 Link Design

SECTION 2 LINK SYSTEM SPECIFICATIONS

An analog optical fiber link is to be designed for the following specifications to carry two analog TV and its associated audio signals, three duplex telephones, and one duplex data channel to the subscriber in an FDM format. Consideration of the uplink arrangement is also to be given.

- Link span length—2 km
- LED transmitter output into 50-μm-core graded-index fiber——10 dBm

- Receiver sensitivity—−23 dBm for SNR of 40 dB
- Fiber loss—3 dB/km
- Fiber bandwidth—100 MHz over 2 km

LINK BUDGET

The link budget is presented in the following table.

Transmitter output	−10 dBm
Receiver sensitivity	−23 dBm for SNR of 40 dB at 30% modulation
Margin	13 dB per channel
Fiber loss 3 dB/km over 2 km	6 dB
Total connector loss 1 dB × 2	2 dB
coupling losses	1 dB
Headroom for temperature, aging, and future splices	2 dB
Total attenuation	11 dB
Excess power margin	2 dB
Fiber dispersion	200 MHz/km

INTERPRETATION OF SECTION 2 FIBER OPTIC LINK SPECIFICATION

The two TV channels, three telephone channels, and one data channel are to be frequency-divisional-multiplexed into a single band of analog signal. It is convenient to place the two TV channels and their associated sound channels by using standard equipment to subcarriers corresponding to two VHF channels, for example, channels 2 and 4. The telephone channels are individually converted to 64-kb/s PCM signals and combined to form a single frequency-shift keying (FSK) channel on a third subcarrier. These carriers can be spaced along the 100-MHz band at positions such that no intermodulation products fall in any of the signal bands. This reduces the linearity requirement of the transmitter greatly at the expense of bandwidth. An automatic-gain-control (AGC) carrier is needed in order to maintain service in case of the failure of any channel.

An uplink is required to provide duplex operation for the telephone and data channels. This link can be provided by using a second fiber or by a wavelength-multiplexed bidirectional transmission along the downlink fiber, if the couplers have negligible insertion loss. This is, in principle, possible since the coupler can be arranged to have relatively large coupling loss to the uplink and almost no loss to the down-

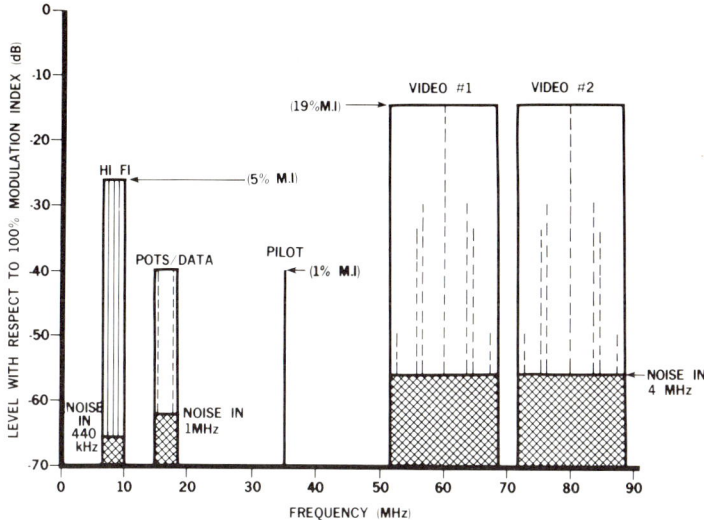

Fig. 9-8 FP2-FP3 Transmission link frequency plan.

link. In this case the digital telephone and data signals are to be sent as a low-bit-rate digital stream. Since the SNR required for a digital signal is over 20 dB less than the analog signals, an acceptable transmission quality is achievable if the uplink has a coupling loss of <20 dB. It is again necessary to point out that the design approach adopted here is not necessarily the most cost-effective. Nevertheless, the wavelength-multiplex method offers a reasonable solution at relatively low cost. The analog signal format is shown in Fig. 9-8.

TRANSMITTER DESIGN

The transmitter consists of an input stage where the analog signals at the appropriate level and modulation depth are combined. This is readily achieved by a simple network. To account for impedance mismatch and level adjustment, the signals are fed to the combiner via attenuators. The combined signal is then amplified to the level suitable for modulating the LED. The LED driver circuit consists of a feedback loop so that the LED can be modulated to near 100 percent depth and still retain a high degree of linearity. The complete transmitter block schematic is as shown in Fig. 9-9. Circuit monitors are used to indicate the presence or absence of signals.

A frequency response of 100 MHz and a relative distortion of ~40 dB for difference frequency and higher-order harmonics can be expected.

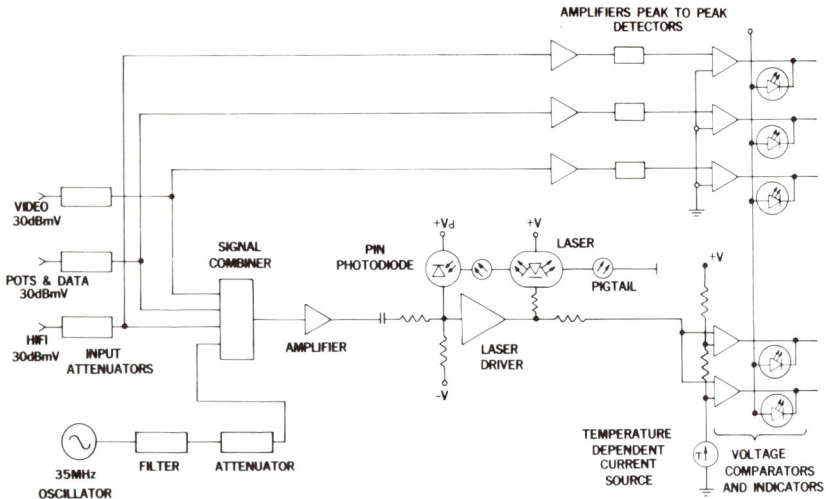

Fig. 9-9 A 100-Mb/s fiber optic transmitter block diagram.

RECEIVER DESIGN

The receiver provides the functions of optical signal detection, amplification, signal separation and provides alarm outputs. A design is shown in Fig. 9-10.

The optoelectronic portion uses a PIN photodiode. The PIN diode has good linearity, generates no excess noise, and is ideal for the high-level signal detection called for in this application, where a high SNR is to be achieved.

The typical amplifier noise level is about 2 pA/Hz$^{1/2}$. The shot-noise contribution of the detector is about the same magnitude. This yields a total noise of about 3 pA/Hz$^{1/2}$. The signal level is −23 dBm. Using a detector diode whose RMS signal current is given by $i_s = 0.5 \, PM/\sqrt{2}$, where P is the average optical power and M is the modulation index, we obtain $i_s = 700$ nA. For the video channel, M is taken as 0.4. This gives an SNR of >40 dB. For the telephone and data channel, M is taken as 0.1, and an SNR of >40 dB is also obtained. This is more than adequate for the PCM signals to be received within 10^9 BER.

The input preamplifier uses a pair of FET transistors in a balanced configuration, and the transimpedance approach is adopted to achieve the low-noise performance. The distortion specification imposes a further design limitation. The feedback resistor of the transimpedance amplifier must be carefully chosen to both achieve the low-noise performance and maintain the distortion level within specification limits.

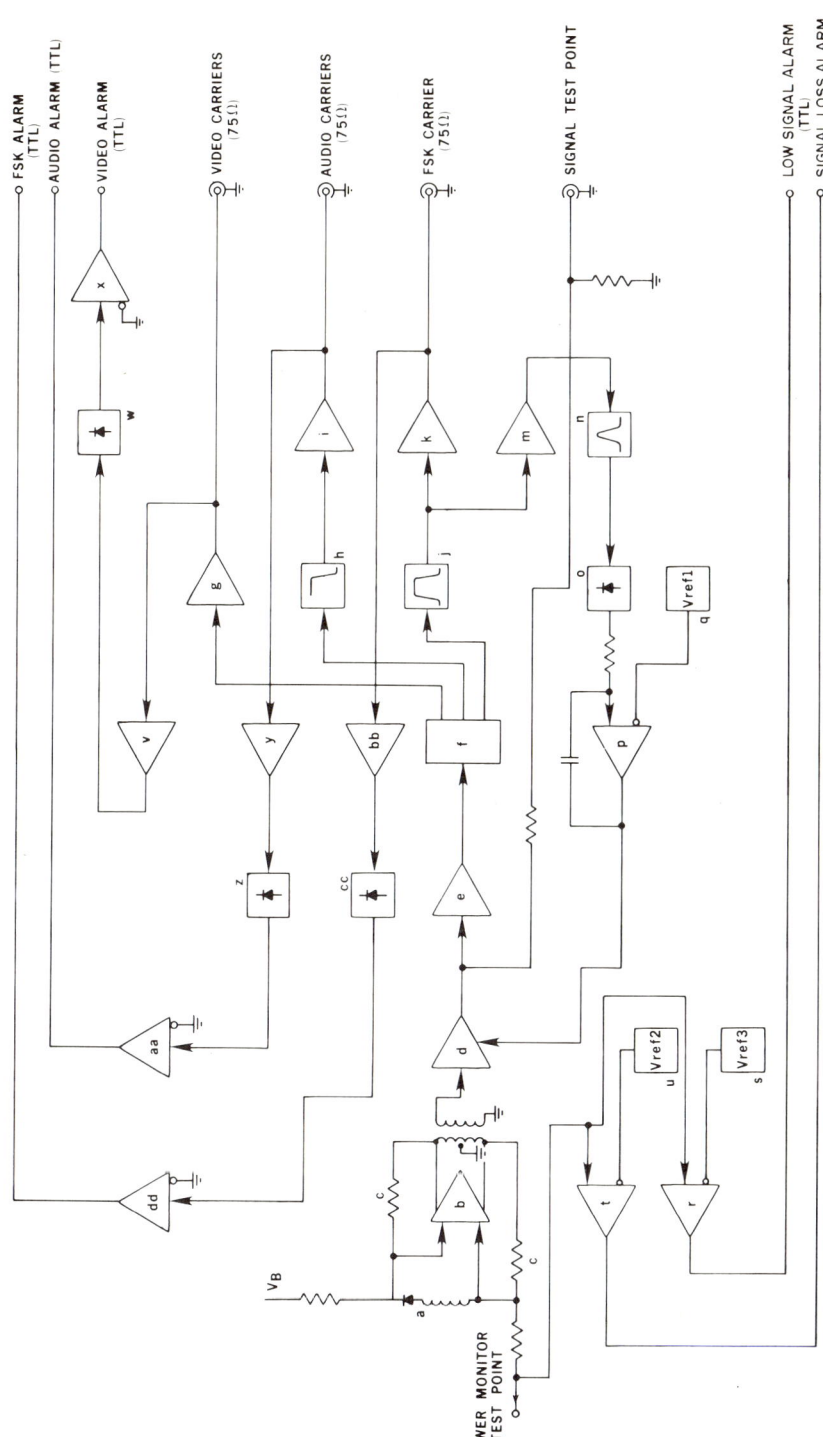

Fig. 9-10 A 100-Mb/s fiber optic receiver.

In the circuit block schematic shown in Fig. 9-10 the PIN detector is designated as *a*, the balanced transimpedance amplifier as *b*, the feedback path as *c*, the controlled gain amplifier which allows adjustment of the signal level as *d*, the buffer amplifier as *e*, and the power splitter as *f*, which achieves signal separation through appropriate filters, designated as *g*, *h*, and *i*. Alarms and monitor points are also indicated and labeled.

The outputs from the receiver for video are appropriate for direct connection to standard TV receivers. The telephone and data signals, however, are in FSK format and must undergo demultiplexing and reconversion from 64-kb/s digital channels to analog form before connecting to the standard telephone and data receiving terminals.

SUBSCRIBER CABLE

The distribution cable to the subscriber premises is assumed to be a two-fiber cable suitable for direct burial form of installation. Furthermore, these basic elemental fiber cable units are to be bundled into a multiunit configuration so that the installation of such cables, designed to cover a neighborhood of several hundred houses, can be practically handled. The unit cable is designed here. One design criterion is to achieve a highly flexible and low-weight structure of small dimension in order that these units can be bundled more conveniently.

A design is as shown in Fig. 9-11. This is a void-filled cable toughened and strengthened for direct burial. The aluminum tape is optional. In many cases it can be dispensed with in order to improve further the cable flexibility.

The designed permissible tension is 120 kg at 1 percent elongation. The void-filling compound can be a petroleum jelly, enabling the cable to be used at subzero temperatures.

LOSS AND BANDWIDTH

The cable for distribution purposes is relatively short. However, in order to meet the power budget, the loss must be as low as possible. This means that the losses associated with connection and splicing must be minimized. Furthermore, the installation procedure is simpler if the cables can be laid without spliced joints. Thus the fiber cables are assumed to be in single unspliced length of up to 2 km. The fiber NA is to be chosen to be as high as possible while maintaining the low basic loss. The high NA allows excess loss to be minimized and permits a smaller and more flexible cable to be designed. The fiber

Anatomy of a Design 181

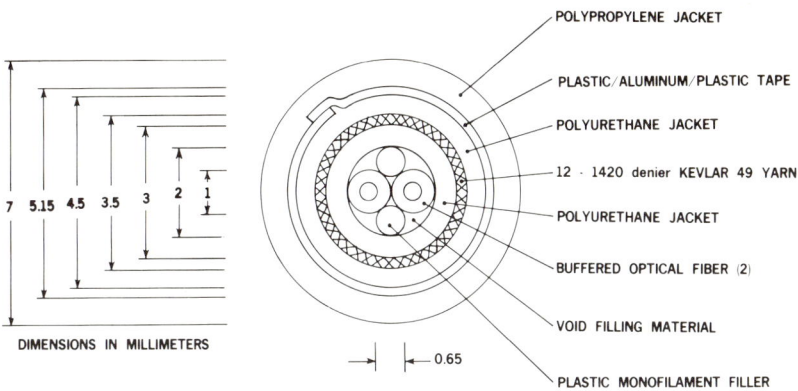

Fig. 9-11 Drop cable.

core size is to be adequately large for minimizing connector loss and increasing coupling efficiency to the source and detector. The core-to-OD ratio, however, must not render the fiber to be excessively sensitive to microbends.

The bandwidth of 200 MHz·km is readily achieved for a relatively high NA, large-core graded-index fiber. The fiber for a distribution cable, therefore, may be designed differently than a fiber for the trunk cable. Since the lengths of fiber needed for distribution applications is large, a separate standard is in order. For this design, a possible fiber with NA = 0.25 core size of 100 μm could be selected, provided that loss and microbend sensitivity are adequate over the operating temperature range.

OTHER DESIGN PARAMETERS

Availability of a system is another design parameter to be addressed in practical systems. The availability data are derived from failure probability analysis of the critical components. In such an analysis, operating conditions and design stress on the components must be defined. The availability, often stated as *percentage time available*, is a derivative of MTBF and average repair time. This requirement impacts on the design principles, not only in terms of types and grades of components and number of such components to be used in a system, but also sets a strict design limit to the safety margin to be maintained. Hence, if a state-of-the-art component is to be used, the operating condition must be made nonexacting; otherwise, the failure probability could rise sharply.

Chapter Nine

COMMENT

The general guideline to system design as outlined here is to first establish the general requirements leading to the specification and then to realize a design for each functional unit. This is a reasonable design approach. It is important to point out that the knowledge of the technological capability of system components helps us to understand how they can be used to solve system problems, and that the system requirements help us to highlight needed technology advances. This design analysis, coupled with the detailed capabilities and limitations of the key components as described in earlier chapters, can be used effectively as a guide to develop other system designs.

Chapter 10

System Economics

The major, and perhaps the only, reason to use a new system product is that it is more economic than existing products for a particular application.[1] A different but also compelling reason for the use of a new system product is that it can provide a new service at an affordable cost. These two reasons are easily stated, but in practice they must be handled carefully. Otherwise, erroneous conclusions may be reached in cases where the comparative reasonings, for or against the new system, are on too narrow a basis or even on an incompatible basis. It is utterly meaningless to state that, since the price of fiber cable is $X + \Delta$ per meter while the price of coaxial cable is X per meter, a fiber system is more expensive than a coaxial system. This statement attempts to deduce from the price of one of the system components the price of a system. Furthermore, nothing much is said about the types of fiber or cable. The conclusion is totally invalid and most misleading. Even a statement to the effect that a certain fiber system designed for a certain information-carrying capacity is more expensive than an equivalent coaxial system may be erroneous, since maintenance, installation, and other operational costs, and terminal equipment price differences could swing the system economics. Other important factors such as system expandability and initial capital outlay could have a very significant effect on the real costs. In fact, a particular economic comparison of competing systems should preferably be carried out with the basis of comparison defined and all pertinent technological and operational factors included. Nevertheless, it is possible for general discussion purposes to construct the economic comparisons in relatively meaningful stages. In the first stage the costs of only the basic equipment necessary to handle information, arranged in a certain signal format from the input of the transmitter to the output of the receiver, are compared. In the second stage the signal-processing cost is added. In the third stage the installation maintenance and expansion capability are included. In the fourth stage other network considerations are taken into consideration. In the fifth stage the affordability of the new service system is evaluated.

First-Stage Considerations

The basis of this stage of economic comparison between optical fiber systems and other solutions is to compare the costs of the basic transmission system equipment consisting of the following parts to their equivalent parts: (1) a transmitter which accepts the input signal to be transmitted in a binary bit stream scrambled or otherwise conditioned for minimum dc content and at the appropriate logic circuit signal level, or in an analog format of certain bandwidth; (2) a receiver which outputs the signal in the same format; and (3) a fiber cable spanning a certain route length without a repeater.

BASIC SYSTEM COMPARISON

When comparisons are to be made, the system bandwidth or bit rate and the number of parallel systems, and the span length must be specified. In comparisons with a copper wire equivalent system, the price of the fiber cable plus the transmitters and receivers is to be compared with the price of the copper cable plus the intermediate repeaters needed along the span length. If the cable costs are the same, the cost of the optical transmitter and receiver must be less than the cost of the intermediate repeaters before a fiber system can have any economic advantage. It is obvious that optical systems become more attractive as information capacity and route length increase, since copper cable and repeater costs increase rapidly with system bandwidth. For short systems involving no repeaters, the optical transmitter-receiver cost will mask the cable cost. In such a case, fiber solution is always more expensive.

Cost trends can be presented graphically to illustrate the cost per unit route length as a function of the system bandwidth or the route length. This is shown in Fig. 10-1. The cost scale is arbitrary. It is intended to provide a qualitative illustration. The step jumps are due to abrupt changes in the cost of the electronic components for different frequency ranges. When comparing microwave or satellite systems with optical fiber systems, the costs of the microwave terminal transmitter and receiver can be matched against the costs of the fiber transmitter

System Economics   185

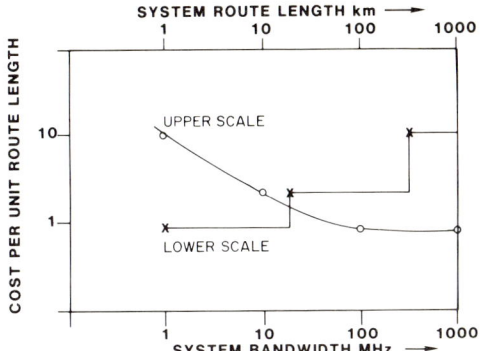

Fig. 10-1 System cost trends.

and receiver. The antennae cost can be matched against the cable cost. In this case it is more difficult to make a general statement of the cost comparison trends for different bandwidths and route lengths. Probably microwave systems are cheaper for long routes and for systems with several parallel channels. This type of comparison is very sensitive to parameters other than the equipment hardware cost.

The stage 1 economic comparison is usually a pessimistic and unfavorable evaluation of the economic competitiveness of the fiber system. This type of comparison should not be used as the basis for discouraging the use of fiber systems. On the other hand, if this type of comparison indicates economic advantage for the fiber system, this advantage more than likely will be even greater when other considerations are given. It is useful as a means for establishing areas of application where fiber systems have overwhelming advantages. Examples of applications of this type could be found in repeaterless fiber submarine cable systems and interoffice trunking at bit rates sufficiently high to require coaxial cable as the transmission medium.

Second-Stage Considerations

The second level of economic consideration takes into account the signal-processing part of the terminal equipment and includes reliability, powering, and supervisory aspects. Effect of the introduction of repeaters is also included.

The signal-processing equipment is included in the economic picture. A fiber system requires almost exactly the same multiplexing equipment as does the other system, except that the equalization of cable characteristics and the fading correction encountered in radio systems are not needed. On the whole, signal-processing equipment for optical systems tends to be somewhat simpler and, therefore, cheaper. For low-capacity systems requiring only the first-level multiplexing, fiber systems tend to be less economical, since the cable cost, particularly the cost of paired copper wire cable of the unscreened type, is low for fiber to compete. For higher-capacity system requiring second-level or higher-level multiplexing, the multiplexing equipment cost predominates over the transmission equipment. Parallel fiber systems have a competitive edge, especially as system capacity and system length increase. This is because the type of copper cable needed tends to have a higher price than the fiber cable, even without the repeater. The cost per unit bandwidth distance of optical fiber systems at different system bandwidths is plotted qualitatively against copper wire systems. This is shown in Fig. 10-2.

For longer fiber systems involving optoelectronic repeaters, the advantage of optical systems at high bit rates is great, while at medium bit rates it is less dominant. This is the result of the relative cable and repeater costs.

It is to be noted that an additional cable cost may be incurred for fiber systems if the repeaters must be remotely powered. However, since the span length of an optical system is relatively long, local power feeding at repeater stations is usually possible.

The costs of providing supervisory facilities and assuring equipment reliability can be high for optical systems. In general, these requirements tend to increase the fiber system cost more than the copper systems. However, since fiber systems have abundant bandwidth, and if error monitoring and supervisory activities are allowed to be carried out along the same transmission line, it is possible to readily design

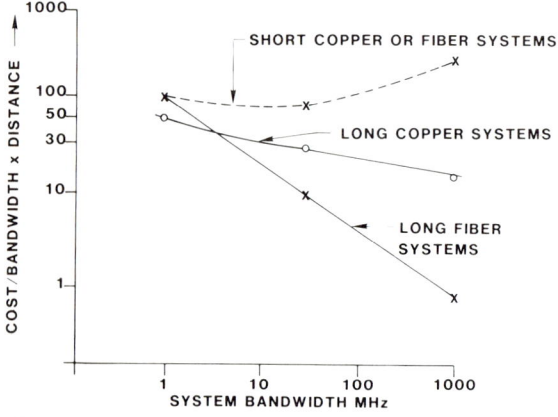

Fig. 10-2 Cost comparison of fiber and copper systems.

in-traffic diagnostic capabilities into the electronics. Otherwise, the need to provide separate order wire will substantially affect the fiber system cost.

Third-Stage Considerations

The third level of economic consideration takes into account more pertinent real-life aspects of a practical operational system. These include installation, maintenance, and operational costs. Cost of money and investment issues in terms of present-value annual cost (PVAC) are also brought into focus.

COST COMPONENTS OF A SYSTEM

The cost of the hardware of a communications system is the basic equipment cost and is the price to a purchaser before installation. For the purchaser to use such a system, a number of other cost components arise. One of these is the cost of installation. Another is the right-of-way cost. Additional building cost also could be a component, even if the main buildings are assumed to be available at no cost. Cost of money is invariably involved, as well as the cost of installation of overcapacity to cater to expected expansion requirements of information traffic over the life of the system. These can generally be regarded as the investment costs. Operational and maintenance costs are additional costs.

INSTALLATION COST

For installation in ducts, by direct burial or on poles, the light weight and the flexibility of optical fiber cables increase the length of cable which can be handled conveniently, result in a reduced number of joints or splices being needed along the route, and permit the use of lighter installation equipment. The installation cost is, therefore, generally lower than that of copper cable. Typically in new ducts, more than 2 km can be envisaged to be pulled in at a time, while 0.5 km is considered long for coaxial cables. In directly buried cable, several kilometers of fiber cable can be readily ploughed in a single continuous length. On poles the lightweight optical fiber cable can be strapped to messenger wires or be designed to be self-supporting without excessive cost. Moreover, the electromagnetic immunity enables the fiber cable to be placed along electrified railway or power-carrying cables. This means that many more installation methods and space can be adapted for fiber cable installations.

Splicing or jointing involves machines and a work force. Fiber cables require fewer splices per unit length but involve somewhat more expensive splicing equipment and lengthy time for making the splices than for the case of copper cables. The average system cost could work out to be almost the same for this aspect of installation.

The terminal equipment installation is unlikely to differ greatly for different cable systems. However, for microwave radio systems, the installation cost involves the construction of antennae and the tower on which the antennae are mounted. The terminal construction costs are considerable. On the other hand, there is no cable installation cost along the route, except to and from the towers and up and down the towers. The installation cost of the microwave radio systems is difficult to generalize. It depends on the system length and terrain configurations.

RIGHT-OF-WAY

When a system is to be installed to connect two centers, it is highly likely that the intervening area between the two centers is not the private property of the system operator. In such a case either a strip of land along the route is to be purchased or a suitable leasing arrangement is to be established.

Along a ducted route the land and usage rights usually reside with the system operating company. Several cables are placed within the duct. The cost of acquiring the land and the installation of the ducts with multiple or single compartments is very high and is a one-time installation cost to be apportioned to current and future cable cost. For accounting purposes, a duct cost is assigned, which is usually proportional to the area of the duct cross-sectional area occupied by the cable. A smaller cross-sectional area cable such as the optical fiber cable, of course, is levied at a smaller cost than a large coaxial cable. This is a significant and dominant factor in favor of the fiber cable. This factor can be so large as to absorb the excess cost of the terminal equipment and prove out a fiber system against a copper cable system in areas where the duct charge is large, such as in city areas. In fact, if duct congestion exists, such as in many major metropolitan areas, the installation cost may be enormous, or installation may simply be impossible other than for the slender fiber cables.

In rural areas where direct buried construction is to be used, acquisition of the right-of-way is necessary. It is possible to bury the line at sufficient depth, in order to permit farming above it. In general, for maintenance purposes, this arrangement is not acceptable. In this type of construction the route length may need to be increased to avoid properties whose owners reserved the right not to grant right-of-way and to avoid other physical obstacles such as deep rivers and highways.

For pole and duct installations of optical fiber cables, right-of-way along highways, railways, and high-voltage electrical transmission lines can usually be leased at a price. The lease cost varies. Optical fiber cables can be installed along any of these routes, since, as mentioned earlier, electrical interference is not a problem. This means that optical fiber systems may find suitable right-of-ways at a cost lower than that in copper systems.

For microwave systems, the right-of-way aspects take on a different situation. So far the irradiation of the airspace between the antennae does not constitute violation of private properties. No right-of-way is exactly involved. The radiowave airspace is controlled by such organizations as the Federal Communications Commission (FCC) in the United States and post, telephone, and telegraph (PTT) systems in European countries through the International Radio Consultative Committee (CCIR). The right-of-way cost is lower than that of the cable systems. However, the land for the intermediate towers must be provided, and the congestion of radio spectrum must be considered. In many major cities the microwave radio spectrum is extremely congested and often not available. Assuming the availability of a spectral window, the right-of-way cost amounts to the cost paid to have the right to use that spectral window. Furthermore, additional cost for routing modification could arise, if tall structures are allowed to be built to block the line-of-sight path.

ADDITIONAL BUILDING COST

This cost includes only the cost of the repeater housing. Most copper cable repeaters are unmanned and are installed in manholes. Fiber cable repeaters can likewise be installed. Since optical fiber systems have fewer repeaters, the cost of repeater housings is lower. However, fiber system repeaters of some types could require ambient temperature control. In that case the additional building cost could be large. This situation is not prevalent since fiber repeater spacing is sufficient to arrange for the repeaters to be housed in stations where drop and insert of signals are needed.

COST OF MONEY

Major system installations must cater especially to the expansion of information traffic expected over the designed life of the system. If the initial installed capacity is equal to the final required capacity, the initial capital investment is obviously larger than if the initial installed capacity caters only to the short-term needs. The cost of money favors the case where initial capacity only is installed first and additional capacity is installed to meet the demand. This is true if the cost of installation of the initial equipment is much lower than the cost of installation of equipment with the finally needed capacity, and if additional equipment can be installed with moderate incremental cost.

In coaxial cable systems the expansion capability is relatively difficult to exploit. A system could be installed initially and equipped to handle x channels of telephones. Because of the change in cable equalization and repeater spacing, the equipment cannot be readily changed to cater to different channel capacities. A system with 34-Mb/s repeaters cannot be easily adapted to carry an 8- or 140-Mb/s signal rate.

In optical fiber cable systems, implementation of the expansion capability is much easier. Since the fiber repeater spacing is almost invariant with signal bit rate and since the fiber bandwidth is adequate for a very wide bandwidth, the changing of bit rate can be accomplished simply by replacing the electronic equipment. Furthermore, each fiber can be designed to work over a spectral range with a wide-bandwidth and low-loss performance adequate for a large variety of system needs. This means that if provisions are made for wavelength selective coupling in and out of the fiber, simultaneous operation at several wavelengths is permissible, thus providing expansion of several times the initial installed capacity without new cable installation.

In microwave systems, with expansion by the insertion of new carriers at different frequencies and polarizations it is possible to increase the initial installed capacity. However, the antenna design allows relatively few carriers to be simultaneously accommodated. Furthermore, the expansion is limited by the total frequency spectrum available.

Comparing the three different approaches the optical system obviously can be designed to be more economic for systems with large expansion requirements on a PVAC.

The importance of using electronic multiplexing and carrier or wavelength multiplexing rather than spatial multiplexing—that is, increasing the number of parallel cables—to increase system capacity is that the cable and installation cost usually is the dominant system cost. Once a cable is laid, it is expensive to lay another cable, and it is also too expensive to lay extra cables. In the case of the microwave system, the installation of extra antennae is relatively easy and economical. However, spectral availability is a fundamental finite limit.

The PVAC is an important basis for economic comparison, especially when the major system cost components have been included. It is prob-

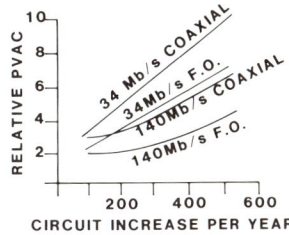

Fig. 10-3 Cost comparison of coaxial and fiber systems for use en route with different circuit increase per year.

ably the only fair method of comparing systems of different channel capacities. It is recommended for use by the International Telephone and Telegraph Consultative Committee (CCITT) as the preferred basis for economic comparisons. In a PVAC calculation the following parameters are assigned with typical values as listed:

Study period—20 years

Circuit growth rate—5–20%

Cost of money—8–15%

Economic life for electronic equipment—15 years

Economic life for cables—30 years

Maintenance—1–2% of capital cost per annum

Provisioning periods for cable—20 years

Provisioning periods for electronics—5 years

Provisioning periods for multiplex—2 years

Typical comparison results for a range of different systems on PVAC basis are shown in Figs. 10-3 and 10-4.

It can be seen that fiber systems tend to be more economical than coaxial systems and that the advantage is more marked for the higher-bit-rate systems. For short-transmission and low-capacity systems, the fiber system is expected to compete eventually with systems using new copper pairs. However, it is easy to see that if an existing copper

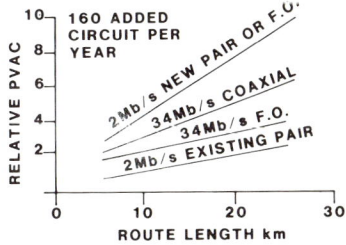

Fig. 10-4 Cost comparison of a 34-Mb/s fiber system with various copper systems.

pair is used, the copper system can be extremely low in cost. Thus it is possible to conclude that for a new network where new cables are to be installed, the advantage of using optical fiber cables is overwhelming. While this conclusion is valid for long transmission systems carrying digital signals at relatively high bit rates, it should not be extrapolated to cover other cases. For distribution systems involving digital or analog signals and short transmission distances, careful economic analysis should be carried out before conclusions are drawn.

MAINTENANCE AND OPERATIONAL COSTS

Operational cost for a modern system involves principally the inventory cost of the spare parts to be maintained. In the early stages, fiber systems incur a higher cost due to the high cost of the key components with relatively short life. This includes laser sources and spare fiber within the cable. As the technology matures, this disadvantage should disappear completely. Fiber and copper cable systems are expected to be comparable in operational costs. The microwave system is not that different and should have operational costs similar to those in other systems.

On the other hand, the maintenance cost of the optical fiber systems is likely to be much lower than that of the copper cable system and is similar to or even marginally lower than that of the microwave system. This is due to the fact that no outside repeaters are necessary for most systems and few repeaters are needed for the long trunk lines. In fact, the long repeater spacing achievable reduces the electronic component count per system down to such a level that transoceanic submarine systems carrying digital signals can be envisaged. This is an important statement of the reliability of optical systems. It implies that the maintenance cost of the optical fiber systems should be low.

Fourth-Stage Considerations

While the third level of economic consideration takes into account a single practical system, the fourth level includes the consideration of related network parameters. The digital fiber trunk system is shown to be economically superior to a copper cable system in most cases, but there has been no discussion as to whether analog implementation is equally or more competitive economically. The influence of the establishment of a national digital network, including electronic exchanges, is yet to be considered. In other words, this fourth-stage consideration takes into account the overall network implications.

TELEPHONE NETWORK

The decision to establish a national digital network resulted from the effort to reduce switching cost. It became obvious that the application of large-scale integrated circuits would drastically reduce the cost of each switching circuit. Moreover, the switch is in fact a special-purpose computer and is capable of many new and important features, allowing the telephone network eventually to expand into a multiservice or integrated service network to handle telephone data and other new revenue-generating services. The impact of introducing digital switching is to alter the cost distribution of network components drastically. First, the interswitching office transmission should be digital. Then the long-haul routes should be made to carry digital signals. Digital subscriber equipment is obviously another logical step toward a wholly digital network.

Within a national digital network the use of analog transmission becomes much less attractive, except on long lines. Even there its competitive edge is rapidly diminishing, especially when optical systems with long repeater spans are introduced. The fewer the number of digital repeaters along the transmission system, the higher the system reliability. This is rapidly compensating for the lower reliability of a digital transmission system, because of the higher component count of its electronic circuit in its repeaters.

Having made the decision to go "all-digital," optical fiber systems economics takes on a new dimension. It now does not impose any economic penalty to the entire network since it is a digital system. In fact, its improved reliability and wide-bandwidth capability will allow the implementation of major transnational digital transmission systems. This happens to be a remaining network requirement toward a national digital network.

TELEVISION NETWORK

The television signals are transmitted and distributed to the subscribers via free space and have lately been supplemented by satellite broadcasting and a coaxial cable network. Further developments are likely toward a switched-TV distribution network capable of handling hundreds of TV channels. While the use of optical fiber for a TV signal transmission network appears to be logical, the analog TV signal, with its requirement for a high SNR and the high cost of converting it into a digital format, are difficulties which must be overcome before a fiber network for TV transmission and distribution is meaningful. The technological development of very high speed digital circuits, a low-cost digital TV signal encoder and decoder, and a TV signal switch could change drastically the economy of an optical fiber TV network.

Fifth-Stage Considerations

Optical fiber systems have many unique characteristics which result in new and improved applications with no comparable alternatives. The economic consideration of these is based on affordability criteria. In a business sense, the affordability issue is linked with perceived needs and other intangibles. From a purely economic standpoint, affordability is associated with the gross national product (GNP) and the influence on other expenditures. As mentioned in Chapter 1 in the discussion of the social impact of communications systems, the optical fiber system has been used in a host of different applications because of its performance and its economic impact. The success of an optical fiber system product is, therefore, dependent on the wisdom and judgment of many people.

A military-base communications system capable of rapid deployment using a minimum of logistic support is an obvious example, where an optical fiber designed for that purpose offers superior performance and overall economic savings. A rapid calculation of transportation and installation costs alone would demonstrate the sound economic base for such a system. The added advantages of a tactical nature are further incentives.

A interference-immune transmission system for an intensive care unit in a hospital, realized with the use of optical fiber, could improve the assurance of administering correct measures to critically ill patients and reduce the load on supervisory personnel. This could result in economic savings in both operational cost and in human life.

To complete the economic considerations of optical systems, a brief look at the economics of a conceptual broadband multiservice network is to be presented. This also serves to illustrate some of the aspects of a basic network of the future as a fitting conclusion to a book devoted to a discussion on optical fiber system technology.

ECONOMIC CONSIDERATIONS OF A BROADBAND LOCAL LOOP SYSTEM FOR MULTISERVICE APPLICATIONS

When a multitude of services are to be provided in the subscriber's premises, the transmission medium is required to have sufficient bandwidth to handle the information, while the terminal equipment is required to have interactive features and operating characteristics compatible with human needs. The system must also have housekeeping functions such as handling instructions, recording billings, and executing commands. All these factors must be implemented in an economically affordable manner as described earlier.

The broader economic issues are:

1. How to justify the installation of a network of optical fiber transmission lines to the subscribers at an early stage when only the telephone service is to be provided along with rudimentary data and TV services.
2. How to encourage the formation of a service industry to provide a host of yet-to-be-defined needed services.
3. How to promote the development of the technology needed to create the subscriber terminal with all the desired characteristics, and to determine what these desired characteristics are.
4. How to encourage investment from public and private sectors toward this endeavor.

To some extent these issues have been considered, and germane ideas have appeared. Development of optical fiber technology, semiconductor device technology, and computer technology are definitely helping. Energy limitations and trends toward an information society also help to highlight the need.

Low-Cost Optical Transmission Line

With respect to the first issue, namely, the construction of the subscriber lines, the solution lies in reducing the fiber cable cost. As the volume of usage increases and as technology improves, there is every hope that fiber cost comparable or lower than a pair of copper wires is attainable.

The massive quantities needed will also allow light source and detector costs to be reduced to the extent that the provision of only a single low-bandwidth service such as the telephone becomes economical.

Information Service Industry

The second issue is the establishment of a service industry based on the vending of information. Experiments concerning the sale of information stored in a computer are already beginning to emerge. For special-interest groups such as airlines, travel agents, and stockbrokers, information-retrieval systems providing timetables, ticketing, stock transactions, and inventory control information are beginning to be available, not only as internally available facilities of a specific organization, but also as commodities available from information vendors. The cable TV industry can be regarded as the vendor of entertainment as opposed to the broadcasters, who are vendors of entertainment but supported by mandatory public dues or advertising revenues.

This industry is expected to blossom into a big and diversified entity, catering to general and more selective interest groups. The availability of an outlet in the form of a broadband multiservice network would accelerate the expansion of this industry.

Related Technologies

This is the pacing issue. Despite the availability of very large scale integration (VLSI) technology, the creation of a universal subscriber terminal with many features needed to appeal to the human users is a challenging task. The development of TV recorders, hard-copy printers, and high-resolution display is important but is only a part of the subscriber equipment needed.

The VLSI technology is still in the early stage of its development. Neither high-speed devices nor large random-access memory chips are available as yet. Their development, however, is progressing. The emergence of 1- to 4-Mbyte memory would enable speech and pattern recognition to be achieved. This would improve the human/machine interface, by "humanizing" the machines. Artificial intelligence is also in its infancy. Computer software evolution has only just begun. Nevertheless, technology is moving in the right direction, toward achieving higher-speed and higher-energy conversion efficiency. If properly directed, the key technologies needed for the universal subscriber terminal should be in place within a decade in large volumes and at an affordable cost.

Investment

This final issue is crucial. The process to encourage investment from public and private sectors is probably evolutionary. The entrepreneurs with their dreams and visions will fit the technology and business jigsaw piece by piece with their efforts, and a completed picture will emerge.

References

1. K. C. Kao and M. E. Collier, "Fibre-Optic Systems in Future Telecommunication Networks," Conference Proc. World Telecommunications Forum (Geneva, October 6–8, 1975), pp 1.3.1.1–1.3.1.6.

Selected Bibliography

- A Special Joint Issue on Optical Electronics with Applied Optics, *Proc. IEEE* **54**(10) (1966).
- M. Born and E. Wolf, *Principles of Optics,* 3d ed., Pergamon Press, New York, 1965.
- R. E. Collin, *Field Theory of Guided Waves* McGraw-Hill, New York, 1960.
- R. H. Doremus, *Glass Science,* Wiley, New York, 1973.
- R. M. Gagliardi, and Sherman Karp, *Optical Communications,* Wiley, New York, 1976.
- D. Gloge, ed., *Optical Fiber Technology,* IEEE Press, New York, 1976.
- M. J. Howes and D. V. Morgan, eds., *Optical Fibre Communications* Devices Circuits and Systems, Wiley, New York, 1980.
- C. Kao, ed., *Optical Fiber Technology II,* IEEE Press, New York, 1981.
- H. Kressel and J. K. Butler, *Semiconductor Lasers and Heterojunction LED's,* Academic Press, New York, 1977.
- D. Marcuse, *Light Transmission Optics,* Van Nostrand Reinhold, Princeton, N. J., 1972.
- D. Marcuse, *Theory of Dielectric Optical Waveguides,* Academic Press, New York, 1974.
- J. E. Midwinter, *Optical Fibers for Transmission,* Wiley, New York, 1979.
- S. E. Miller and A. G. Chynoweth, eds., *Optical Fiber Telecommunications,* Academic Press, New York, 1979.
- W. K. Pratt, *Laser Communication Systems* Wiley, New York, 1969.
- C. P. Sandbank, ed., *Optical Fibre Communication Systems,* John Wiley, New York, 1980.
- Special Issue on Fiber Optics, *Transact. IEEE* **Com-26**(7) (1978).
- Special Issue on Optical Communication, *Proc. IEEE* **58**(10) (1970).
- Special Issue on Optical Fiber Communications, *Proc. IEEE* **68**(10) (1980).
- Special Issue on Rays and Beams, *Proc. IEEE* **62**(11) (1974).
- H. G. Unger, *Planar Optical Waveguides and Fibers* Clarendon Press, Oxford, 1977.

Index

Absorption edge, 11
Acoustic sensor, 157
Analog-to-digital (A/D) conversion, 165, 175
Attenuation measurements, 56–57
    insertion loss, 56
    optical time domain reflectometry, 57
Audio receiver, 174
Automatic gain control (AGC), 176
Avalanche photodiode (APD), 105, 110, 112, 113, 115, 140

Bandwidth (pulse dispersion), 11
Bandwidth analysis, 17
Bend test for mechanical evaluation of cable, 81
Bessel function, 27, 28
Bounded (nonradiative) propagation, 24
Business development applications of fiber systems, 1, 4–5, 7

Cable(s), 75–82
    distribution, 155, 180
    tow (tether), 150
    undersea, 140–141
Cable deployment, 147
Cable design, 146, 148–149
Cable strength members, 76–78
    carbon fibers, 77, 78
    glass fibers, 77, 78

Cable strength members (*Cont.*):
    plastic monofilaments, 76, 78
    steel wire, 76, 78
    textile fibers, 76–78
Cable structures, 77, 79–81
    loose and tight configurations of, 79, 80
Cable television (CATV), 8, 143
    CCTV (closed-circuit TV), 9
    costs of, 193–194
    design considerations of, 156
    switched TV network, 8
    trunking, 143–144
Cable testing, 80–82
Characteristics of fiber waveguide:
    attenuation (loss), 11
        microbending, 41, 42
        versus wavelength, 12
    bandwidth (pulse dispersion), 11
    dispersion, 13, 31
    immunity: to electromagnetic interferences, 3, 13
        to fire hazard, 13
        to radiation, 13
    loss (attenuation), 11
    material dispersion, 31, 87
    measurements of mechanical, 59–62
    measurements of optical, 56–59
    optical, 11–13
    physical, 10–11
    profile dispersion, 87
    pulse dispersion (bandwidth), 11

Characteristics of fiber waveguide
    (Cont.):
  special, 13–14
  temperature limit, 13
  waveguide delay, 31
Communications network, fiber systems
    in, 3–5
  advantages of, 4
  costs of, 193–194
  evolution of, 4
Connectors, 125–128
  ferrule, 126
  joint loss of, 118
  lens, 128
  military applications of, 146
  precision butt joint, 125
  precision transfer-molded single fiber,
    126–127
Copper wires, 10
Cost(s), 183–196
  installation, 187–188
  light source, 89
  maintenance and operational, 192
  present-value annual (PVAC), 187, 190,
    191
  right-of-way, 188–189
  telephone network, 193
  television network, 193–194
Couplers, 129–134
  area-splitting, 131–132
  beam-splitter, 132–133
  bidirectional dichroic, 133, 149
  diffusion, 129–131
  evanescent field, 130
  reflective star, 131
  transmission star, 131, 132
  wavelength-selective, 133–134
Critical angle, 23

Delay distortion, 57
  direct measurement of the transfer
    function, 58
  single-pass impulse response
    measurements, 57–58
Demodulation, 103
Digital audio transmitter, 167–169
Digital-to-analog (D/A) conversion,
    173–175
Digital video transmitter, 165–167
Dispersion, 13, 31
Distortion (see Noise)

Distribution cable, 155, 180
Distribution systems, 138, 150–151,
    156
Double crucible fiber-drawing technique,
    67, 69–70

Educational materials from fiber systems,
    1–2, 8–9
Emitter-coupled logic (ECL) levels, 173,
    174
External chemical vapor deposition of
    glass (plasma CVD) process of fiber
    fabrication, 66
External chemical vapor deposition of
    soot (external CVD) process of fiber
    fabrication, 63–64

Fiber, 17–18, 21–73
  chemical vapor deposition of, 63–66
  coating of: silicone room-temperature
    vulcanization (RTV), 71
    ultraviolet (UV)-cured epoxy, 71
  connecting of (see Connectors;
    Couplers)
  dimensional tolerances in connection
    of, 118
  elliptical, 40
  end preparation of, 120–123
  evaluation methods for, 55–56
  fatigue of, 21
    (See also Glass fatigue)
  formation of: double crucible, 67,
    69–70
    flame hydrolysis, 63
    phasil system, 68–69
  graded-index, 21, 22, 35–36
    bandwidth versus wavelength of,
      36
  large-core step index, 21
  minimum lifetime of, 55
  misalignment losses in connection of,
    119–120
  multimode step-index, 34–35
    cladding thickness of, 34
  physical structure of, 22
  plastic-clad, 36–37
    radiation resistance of, 37
  polarization maintaining in, 157
  single-mode, 21, 32–34
    lossy outer coat of, 33

Fiber (Cont.):
  sintering of, 63
  splicing of (see Splicing of fibers)
  step-index, 22, 34–35
  (See also entries beginning with term: Glass)
Fiber cables (see entries beginning with term: Cable)
Fiber casting, 71
Fiber core and cladding, 21
  birefringence of, 40
  diameter variations of, 38–39
  eccentricity of, 40
  ellipticity of, 39–40
Fiber-drawing processes, 69–71
  double crucible, 67, 69–70
  from fiber preforms, 70–71
Fiber durability, 52–54, 170–172
Fiber-fabrication processes, 62–69
Fiber imperfections, 21, 37–41
  and basic material properties, 37–38
  in bulk of material, 38
  dimensional fluctuations in fabrication process, 38–41
    birefringence, 40
    diameter variations, 38–39
    eccentricity, 40
    ellipticity, 39–40
  effect of, 40
  externally induced, 40–41
  microbending, 41, 42
Fiber loss and dispersion, 170
Fiber packaging, 41–44
  elastic deformations and, 42, 44
  protective jackets for optical fibers, 43
Fiber profile, 59
Fiber strength, 51–52, 170–172
  and larger flaws, 51
  measurement of, 59–62
  and microcracks, 51
Fiber system(s):
  application of (see Services performed by fiber systems)
  distribution, 138, 150–151, 156
  economics of (see Costs)
  sensor, 138, 157
  specifications for, 159–182
  transmission, 137–138
Fiber-to-detector junction, 18
Fiber-to-fiber junction, 18
Fiber unit design, 169–170

Fiber waveguide (see Characteristics of fiber waveguide; Propagation in waveguides)
Field-effect transistor (FET) amplifier, 115, 178
Fourier transform, 12, 58
Frequency-division multiplex (FDM), 104, 160, 161, 175, 180
Frequency response of photodetector, 107
Frequency-shift keying (FSK), 104, 176
Fungus testing for environmental evaluation of cable, 81
Furnaces, 70

Glass:
  flawless, 45
  imperfections of, 37, 45
    crystallization, 51
    liquid state, 38
    phase separation, 51
    scattering, 38
    (See also Fiber imperfections)
  inorganic, 10, 13, 37
  organic polymer, 37
  strength of, 44–45
    (See also Fiber strength)
  surface structure of, 47–48
  theories of strength of, 45–46
  types of, 10–11, 13
  (See also entries beginning with term: Fiber)
Glass fatigue, 21, 48
  crack propagation in, 48–51
  cyclic fatigue test, 50
  dynamic fatigue test, 60
  fiber (see Fiber strength)
  minimum lifetime in, 55
  static fatigue test, 50, 61
Glass with flaws, 45–46
  stress-induced, 45
Graphite furnace, 70
Griffith equation, 46

Hackle zones in glass fracture, 50, 121
Hankel function, 27, 28
Health care applications of fiber systems, 1, 5–6
High-impedance front-end amplifier design technique, 114–115
Human/machine interface, 196

Index

Impact resistance test for mechanical evaluation of cable, 81
Information carrier, 10
Information services, 4
Information society, 1
Infrared (IR) absorption edge, 11
Insertion loss, 56
Internal chemical vapor deposition of glass (internal CVD) process of fiber fabrication, 64–66
Internal reflection (see Total internal reflection)

Lasers, 85
 (See also Light sources)
Lasing threshold, 93, 96
Light-emitting diode (LED), 17, 86, 98–99, 101, 104, 140
Light-source-to-fiber junction, 16–17
Light sources, 9, 10, 83–102
 arc lamps, 84–85
 coherence of, 89
 coherent, definition of, 83
 cost of, 89
 gas discharge, 84–85
 gas laser, 17, 85
 incandescent lamps, 84
 incoherent, definition of, 83
 laser emission: spontaneous, 85
  stimulated, 85
 LED (light-emitting diode), 17, 86
 life of, 87
 modulation of, 88
 operating temperature of, 88–89
 physical size of, 86
 selection of, 90
 semiconductor (see Semiconductor light source)
 solid state laser, 85
 spectral emission of, 87–88
 system characteristics of, 16, 86
Link margin, 16, 164, 168

Maxwell electromagnetic wave theory, 31
Maxwell equations, 22, 27
Mean time before failure (MTBF), 87, 181
Measurements:
 bandwidth, 58
 fiber strength, 59–62
 impulse response, 57–58

Measurements (Cont.):
 insertion loss, 56
 of mechanical characteristics of fiber waveguide, 59–62
 of optical characteristics of fiber waveguide, 56–59
 refractive index profile, 59
 time domain reflectometry, 57
Microbending, 41, 42
Military defense applications of fiber systems, 1, 6–7, 137–138, 145–149
Minority carriers, 90, 91
Mirror zone in glass fracture, 50, 121–122
Mist zone in glass fracture, 50, 121
Moisture resistance test for environmental evaluation of cable, 81

Noise, 105
 dark current and signal-dependent, 108
 laser, 89
 laser modal, 105
 receiver, 106
 (See also Signal-to-noise ratio)

Office of the future, 7, 151
OH^- ions, 11
 resonance frequency of, 11
 spectral loss curve of, 11
Operating temperature of light sources, 88–89
Oxyhydrogen flame, 70

p-n junction, 90, 91, 94, 98, 99, 110
Phase-shift keying (PSK), 104
Phasil system of fiber fabrication, 68–69
Phonon, 11
 absorption edge of, 11
Photodetectors, 9–10, 18, 107–112
 avalanche action in, 110
 photoconductive devices, 109
 photoemissive devices, 108–109
 photovoltaic devices, 110–112
 properties of, 107–108
Physical imperfections (see Fiber imperfections; Glass, imperfections of)
Plane wave propagation, 26
Plane wave in a stratified medium, 25

Plasma:
  heterogeneous, 66
  isothermal, 65
Positive-intrinsic-negative (PIN)
    photodiode, 110, 113, 115, 140
Power budget:
  example of, 19
  preparation of, 16
Power efficiency of light source, 86–87
Power output of light source, 86
Preform profile, 59
Profile measurement, 58
Proof test, 62
Propagation in waveguides, 22, 26–31
  approximate eigenfunctions, 29
  eigenfunction of, 27
  electromagnetic wave formation of, 27
  geometric optics description of, 21–22
  meridional ray, 26
  modal description of, 28–29
  plane wave, 26
  skew ray, 26, 27
Pulse-code modulation (PCM), 137, 140, 162, 163
Pulse dispersion (bandwidth), 11

Quantum efficiency of photodetector, 107
Quantum limit, 114
Quasi-Fermi energy levels, 93

Radar remote, 147
Rayleigh scattering law, 38, 57
Receiver, 9, 18
  audio, 174
  design of, 142, 144, 149, 178–180
  high-impedance front-end amplifier design, technique for, 114–115
  noise level of, 106
  perfect, 112
  transimpedance approach to design of, 115
  video, 173–174
Reflection coefficients, 24–25
Reflection at a dielectric interface, 22–24
Refraction at a dielectric interface, 22–24

Satellite ground stations, entrance links in, 139–140

Semiconductor light source, 16, 90–102
  bandgaps and lattice spacing of, 90–92
  double-heterojunction device, 94
  feedback control of, 96
  filamentary emission zone of, 97
  GaAlAs laser characteristics, 95–96
  GaAlAs laser structure, 93–95
  GaAlAs LED structure, 98–99
    edge-emitting, 98
    surface-emitting, 99
  InGaAsP material system, 101
  optical characteristics of, 97–98
  photoemission from, 90–93
  radiative lifetime of, 91
  radiative recombination, 90
  reliability of, 100–101
Semiconductor p-n junction, 90, 91, 94, 98, 99, 110
Sensor systems, 138, 157
Services performed by fiber systems:
  business development, 1, 4–5, 7
  communications, 1, 3–5
  community, 1, 6
  education, 1–2, 8–9
  energy resource management, 1
  entertainment, 1, 7–8
  health care, 1, 5–6
  military defense, 1, 6–7, 137–138, 145–149
  transportation, 1, 3
Signal detection, 106–107
Signal-to-noise ratio (SNR), 57, 105–107, 114, 115, 139, 144
Silanol (SiOH), 47–48
Silicone room-temperature vulcanization (RTV), 71
Single-mode waveguide, 32
Sinusoidally modulated continuous-wave (CW) light signal, 58
Snell's law, 23, 24
Spectral emission of light source, 87–88
Spectral response of photodetector, 107, 108
Splicing of fibers, 18, 123–125
  fusion, 123–124
  grooved substrates, 125
  housing for, 172–173
  nonfused, 124
Stratified medium, 25

Telephone network:
  costs of, 193
  interoffice link, 138–139
Temperature cycling test for
    environmental evaluation of cable,
    81
Tensile strength test for mechanical
    evaluation of cable, 81
Time-division multiplex (TDM), 162
Total internal reflection, 21, 23
  Brewster angle of, 25
  coefficient of, 24
  Snell's law of, 23, 24
Tow (tether) cable, 150
Transfer function, 17
  direct measurement of, 58
  typical values, 19
Transistor-transistor logic (TTL), 174
Transition ions, 11
Transmission:
  analog signal, 162, 163
  digital signal, 162, 163
Transmission coefficients, 24–25
Transmission line, 9, 10
Transmission systems, 137–138
Transmitter, 9
  design of, 141, 145, 149, 177
  digital audio, 167–169
  digital video, 165–167
Transportation applications of fiber
    systems, 1, 3

Trunking in cable TV, 143–144
24 or 30 channel PCM link for a high-
    electromagnetic-radiation field
    environment, 140
Twist test for mechanical evaluation of
    cable, 81

Ultraviolet (UV)-cured epoxy, 71
Ultraviolet (UV) electron absorption edge,
    11
Undersea cables, 140–141

Very large scale integration (VLSI)
    technology, 196
Video receiver, 173–174

Wavelength multiplexing, 88
Weapon guidance, 147–148
Wired city, 6, 7, 138, 151, 155
Wired office, 7, 151

Young's modulus, 46, 76

Zirconium furnace, 70

About the Author

Charles Kuen Kao is Vice President and Director of Engineering for the ITT Electro-Optical Products Division. A pioneer in the field of optical fiber communication, his experience includes theoretical studies and basic research in optical communication systems, fiber optic waveguide communications, overmode waveguide systems, circuits and systems design, and quasi-optical techniques applicable to microwave systems. Dr. Kao received his Ph.D. in electrical engineering from the University of London in 1965. He holds 20 patents (another 9 are pending) — has received numerous awards — has since 1975 delivered some 25 papers at conferences — and has published more than 40 technical articles.

A